TAPE-RECORDING FROM A TO Z

Tape Recording from A to Z

by

Doug Crawford

KAYE & WARD · LONDON

A. S. BARNES AND COMPANY
NEW YORK AND SOUTH BRUNSWICK

First published by
Kaye & Ward Ltd
1974

First American edition published 1975
by A. S. Barnes and Company, Inc.,

ISBN 0 7182 0979 6 (Great Britain)

ISBN 0 498 01680 3 (USA)

Library of Congress
Catalog Card No. 74-17780 (USA)

All enquiries and requests relevant to this title should be sent to the publisher, Kaye & Ward Ltd, 21 New Street, London EC2M 4NT, or A.S. Barnes and Co. Inc., Cranbury, New Jersey, and not to the printer.

Printed in Great Britain by
Biddles Ltd, Guildford, Surrey.

Contents

SECTION ONE

How tape-recording works 1

How to operate your tape-recorder 7

Your choice of tape-recorder 9

SECTION TWO

An alphabetical list of tape-recording terms, techniques, equipment, effects and operation 15

SECTION THREE

The cassette recorder: its uses, how it operates, maintenance, allied equipment 91

The tape cassette: how it operates, various types available, their advantages and disadvantages 102

Chromium dioxide tape 108

The Dolby system 109

The radio studio: its equipment and uses 111

Section One

How tape-recording works

How to operate your tape-recorder

Your choice of tape-recorder

How Tape-Recording Works

Before starting out to learn just how to get the best out of your tape-recorder, it is a good idea to spend a few minutes examining in a non-technical way how exactly tape-recording works.

All of us are aware that the living body possesses sensory organs that flash data to our central control mechanism and so make us aware of the sights and sounds around us. Sound sensations are conveyed to the brain (which then recognises and codes them like signals) as a result of a vibration of the ear-drum, which reacts to increasing and decreasing pressures upon it. These vibrating pressures are called *sound-waves* and usually operate in a medium such as air for land-bound creatures. But you will know that it is possible also to detect sound-waves while, for example, submerged under water. Whatever source is producing the sound it expends energy in compressing the medium around it (air, water) and the vibrations thus produced spread out as a wave in all directions. Using the ear, the brain decodes the sound-wave, measuring its loudness and assessing how rapidly it is vibrating. A perfect ear can detect sound-waves that vibrate as low as 25 times per second and as high as 20,000 times per second. The rippling of the sound-waves affects the ear-drum and the bones and fluids inside the ear, and these movements in turn trigger off nerve impulses which travel to the brain. It is these impulses that are 'heard' by the brain and can be regarded as symbols of the sounds heard rather than the sounds themselves. Sound *frequencies* or rates of vibration are referred to as cycles per second (also known as Hertz). High frequencies produce a high pitch experience in the brain, while low frequencies produce a low pitch. In practice we recognise that the sound-waves arriving at our ears are a very complex mixture of different frequencies and changing volumes, and it is the brain's assessment of these that gives us the various experiences of speech, music, etc. A problem arises when we decide we want to keep a particular set of sound-waves 'bottled up', so that at a later date we can re-experience a particular set of sound vibrations—a bird song, a pop tune or baby's first cries. We must therefore find a system to convert the waves into permanent form and, as it were, freeze them where they stand.

One of the earliest efforts to do this was the Edison phonograph. This consisted of a speaking trumpet which directed sound-waves on to a simple diaphragm in the middle of which was fixed a needle. The needle pressed on to a wax cylinder, which rotated with the aid of a clockwork mechanism that at the same time drew the needle gradually along the length of the cylinder as the recording progressed. The user would speak into the trumpet, causing the diaphragm and needle to vibrate in sympathy with the sound-waves. The needle cut a wavy groove into the rotating wax cylinder faithfully converting the voice into a frozen recorded form. To play back the 'record', the needle was simply run down the groove and the contours would cause a vibration of the needle and diaphragm—which in turn compressed and de-compressed the air around it, re-creating the sounds that had been recorded.

The modern record-player represents a number of advances on this early technique. Today discs are press-stamped from a master mould, but the principle of the wax record is little different from Edison's cylinder: both are mechanical recordings of sound-waves. In place of the old speaking trumpet we use a microphone placed near the speaker's mouth. The microphone converts the sound-waves striking it into electrical impulses which are

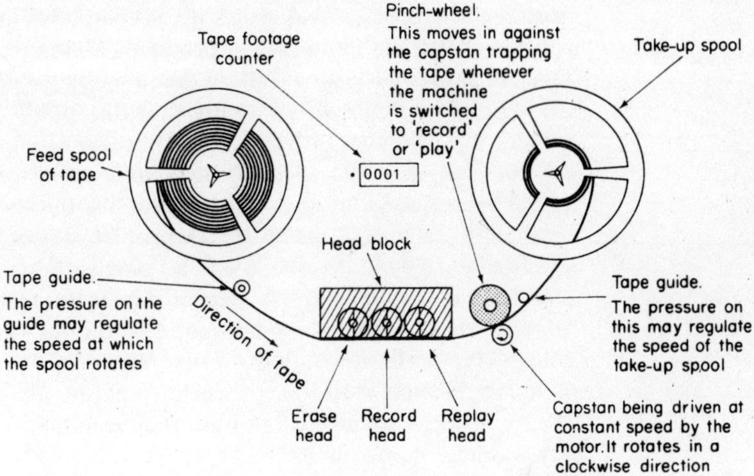

The record and replay heads may be combined into one head

FIGURE 1—THE TAPE DECK

vibrating in sympathy with the received sound. This electrical signal can easily be amplified and applied to the master record cutter, where the signals are converted into mechanical vibrations which in turn cut the wavy groove into the surface of the master disc.

Electronics are used again to play back records, the amplifier and loudspeaker producing a greater volume of more accurately controlled sound than the old needle and horn system.

Although it is comparatively easy to slip a disc on to the turntable and play it back, the manufacture of records is not an easy process. There are problems with the acetate waste that is cut out of the disc which has to be vacuum sucked away to avoid fouling the cutter head. There is no possibility either of removing any mistakes made in the recording. There are also too many energy conversions—the sound into electronic impulses, these into mechanical vibrations which finally cut the record.

A simpler method would be to store the electronic signals in some sort of memory bank to be reconverted into sound at will. Such a system uses a length of tape that is coated with magnetic ferric oxide: and this system of storing sound is therefore known as *tape-recording*.

We will be commenting on the various parts that go to make up a tape-recorder in the A to Z section of this book, but basically a tape-recorder consists of the following components: a motor system that pulls magnetic tape at a constant speed past the recording heads, and winds it safely on to a spool (or back into a cassette or cartridge). The recording heads consist of small electro-magnets—coils of wire that produce magnetic fields in sympathy with the electronic signals passed through them.

At any given time the minute section of tape that is positioned immediately in front of the head will be magnetised by the magnetic field which at that precise

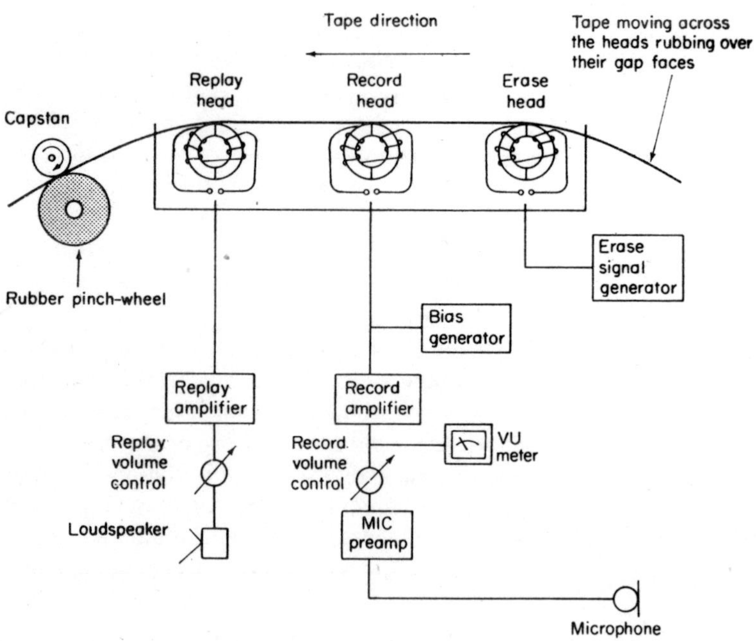

FIGURE 2—THE TAPE-RECORDER CIRCUIT

moment is being produced in the recording head. Seconds later the tape will have moved and a new section will be positioned in front of the head, to be in turn magnetised by the new magnetic field produced by the changing signal.

In this way, as the tape passes the head, it receives a track of varying magnetisation which alters *in sympathy* with the sound that has been fed into the recorder's microphone. On replay, the recorded tape is run at the same speed past the replay head. The changing magnetic fields in the tape produce in the coil of the head sympathetic electronic signals. These currents are then boosted by the amplifier and when applied to a loudspeaker will faithfully reproduce the sounds that were originally recorded.

How to Operate your Tape-Recorder

Your tape-machine will have two mountings on its deck if it is a normal reel-to-reel recorder. The mounting on the left is to take the full spool of tape, and that on the right of the deck for the take-up spool. Between these two mountings is a head block and to the right of this the capstan and pinch-wheel. When the machine is switched to record or play the capstan and pinch-wheel close in on the tape and pull it across the heads and on to the pick-up spool. When the machine is switched to 'spool', 'wind on' or 'wind back' the two spool mountings rotate at speed so that the tape can be rewound or spooled on to the take-up reel.

The tape must be correctly laced across the deck around any tape guides, across the heads, between the capstan and pinch-wheel and on to the other spool. The tape guides differ from machine to machine, and some recorders have guides that vary the tension on the feed spool according to the tape pressure on the guide. It is best to read the manufacturer's booklet to make sure your tape lace-up is correct. Many home machines have a concealed head block and the tape must be slotted into the plastic housing around the heads from above. Cassette machines are basically the same as a reel-to-reel recorder but they are laced-up automatically by dropping in the cassette.

With your recorder switched on at the mains, or with new batteries fitted if it is a portable, it can be set in motion by switching to 'play' or 'record'. The machine usually has its own loudspeaker, and for playback all that is required is the volume control turned up to a comfortable level.

For successful recording there are a number of considerations: whether the sound arriving at the microphone is the sound you really wish to record; whether the type of microphone and its placing will pick up the sound faithfully; has the record volume control been set correctly; and have the machine's heads and tape guides been cleaned. These topics are expounded in the A to Z portion of this book under the headings **Acoustics, Microphones, Record** and **Your Ear and The Microphone.**

Your recorder will have one or more inputs to receive the microphone, or a lead from the radio or disc player. One of these sources must be connected and the record volume control adjusted properly before the machine is switched to 'record'. The step-by-step procedure for adjusting the machine is described under the heading **Record** later in the book.

Some cheaper machines, particularly cassette recorders, have no record volume control, but adjust their recording level automatically. Other machines can be switched from manual control to automatic at the flick of a switch. The reasons why this box-of-tricks is not as clever as it appears are discussed under the headings **Autorecord** and **Limiter.**

Tape-recording is easy. I hope that by consulting the appropriate section, in the A to Z portion of this book, on any topic of recording, before your throw the record switch, that your tapes will become excellent.

Your Choice of Tape-Recorder

There is such a wide variety of machines available on the market that your choice will depend upon a number of factors, not least of which will be the uses to which you intend to put the machine, and the price you can afford to pay.

The large mains-operated tape-recorders are suitable for indoor recording, editing and quality playback. They are ideal for the enthusiast and perform any task that does not demand portability. Sophisticated recording and portability is offered by a number of reporter's machines. These will run from batteries or can be used through a mains adaptor unit. They are ideal for radio interviewers to record their location material, but only the very expensive models are easy to edit on and most are only capable of taking five-inch spools of tape.

Cassette machines are cheap and portable. They are particularly suitable for outdoor work when you do not require to edit the tape. Even when editing is necessary, the cassette can always be copied to ordinary tape and edited later.

There are recording machines that are really items of office equipment and are used for recording notes, letters, messages and memos by the boss, and for playing them back by the secretary who has to type them. These dictating machines have of necessity to incorporate a short playback mechanism, so that the last sentence or so can be listened to before continuing. For transcribing the tape the operator needs a stop/start switch, often foot operated; a short playback switch; and a control to reduce or increase the sound output from the machine.

If you wish to produce different types of artistic sound tapes, your choice should be a large mains operated machine. If a cheap cassette recorder is also bought then this can be used on location, and for adding sound effects or interviews to a master tape being recorded on the main machine. Your main recorder should be able to take at least seven-inch spools, and should preferably have an open head block, so that you can see the tape passing the heads and be able to mark the tape on the replay head for editing. Your recorder may be in mono or stereo, and that choice is only one that you can take, but today more and more machines are being manufactured as stereo models only. You may buy equipment that records twin track or four track tapes. These terms are fully explained under their own headings in the A to Z section of the book. Four track machines are only useful if you wish to conserve tape, for example if the recorder has been bought simply to provide background music. A twin track machine is more suitable for producing creative tapes.

For serious work you will be interested in a recorder with variable tape speeds. Tape speed is measured in inches per second, or centimetres per second, and the uses of the various speeds are explained under the heading **Speed** later in the book. There are some machines with a built-in sound mixer next to the tape deck which is an excellent facility for the enthusiast. For simply recording and playing back pop music, or as a second machine, the cassette recorder is excellent. A lot of recorded commercial music is now available on cassette as well as LP records so that a good cassette machine can easily replace the record-player. The **Cartridge** system of recording is similar to the cassette but involves a different way of winding the tape. It is used for replaying commercial music recordings on eight-track cartridges and is frequently fitted in cars.

Other Considerations

Several refinements are incorporated into tape-recorders and it is important to check that your choice of machine has these facilities.

They include:

A recording indicator, known as the VU meter, to ensure that the recorded signal is being kept above and below certain limits to produce a good recording.

A record volume control to adjust the recording signal. Some cheap recorders simply record automatically, but do not reproduce the original sound faithfully.

A facility to record direct from a radio or record-player and to play back through a separate loudspeaker.

A replay amplifier of sufficient power to drive an extension speaker. See heading **Amplifiers.**

A facility for monitoring through headphones as you record. This is particularly useful if you wish to record sounds from various sources and mix them on to the tape.

If the recorder is to be used for production work, it should have open heads for ease of editing and a selection of tape speeds.

Finally examine the machine carefully before you buy it to check that its construction is robust enough to cope with constant handling. Particularly pay attention to the deck switch that puts the machine into the play, record and rewind modes. This switch must be capable of a long life since it will receive the most wear.

Section Two

An alphabetical list of tape-recording terms, techniques, equipment, effects and operation

Acoustics

The acoustic properties of a room in which a sound is recorded will affect how it actually sounds. Wallpapered or painted walls will directly reflect any sound-waves hitting them and the waves arriving at the microphone in a 'hard wall' room will consist of the original sound mixed with these reflections which arrive slightly later. This results in an echoey effect called *reverberation*. Reverberation can become intolerable in a large hall with plenty of hard surfaces and no soft furnishings to absorb the many reflections. Try giving a speech in a canteen!

Reverberation is controlled by changing the surfaces in a room to be used for recording, and by introducing barriers to the reflected sound-waves. The first necessity in the room is a carpet, the bigger and thicker the better. This will eliminate reflections from the floor, and absorb waves reflected downwards by the ceiling. Tapestry or curtaining over the walls will stop lateral sound-waves from bouncing around the room, and soft furnishings, even human bodies, will absorb the sound-waves.

If a recording room is lavishly treated with these sound dampening materials any speech or music recorded will appear to the ear 'dead', i.e. without any reverberation. This can be very useful if an out-of-doors effect is required as the actor's recorded voice has no roomy atmosphere and can be simply mixed with a sound effect of wind or street noises. But for most purposes a completely 'dead' room without any reverberation will not produce an attractive sound. Acoustic treatment should be limited to say two walls to produce a more lively atmosphere. Music requires more reverberation than speech, so if the room is not very large and there are a few absorbent musicians present try rolling up the carpet to liven the sound.

On many occasions it may seem impossible to change the acoustics for a recording: for example during a short radio interview the reporter cannot start refurbishing the interviewee's office. But if you consider that a good acoustical sound is one that contains few reflections it is possible to improvise. Deaden the room by pulling the curtains, not forgetting to close the windows first. Place the microphone near to the speaker or the musical instrument so that the highest proportion of sound reaching it is direct and not reflected from the walls. The microphone should never be placed on or above a desk or table as this will provide a reflecting surface very close to the mic. The tables used in radio studios have a special non-reflecting surface to avoid this. If a speaker is being recorded, he should be in the softest part of the room, maybe sitting on the sofa rather than at his desk, and preferably facing one of the room's corners so that his voice is not reflected straight back. If he speaks softly to a close microphone there will be less reverberation than a loud voice bouncing around the room to a distant microphone.

To judge the acoustics of a room with the microphone in a certain position it is best to carry out a test recording, but this recording must not be played back in the same room as it will colour the sound again as you listen. Take the tape-machine into a soft lounge, or out-of-doors, and listen carefully to the recording. If it sounds like a tape-recording and not the real thing then it is a bad recording either because the recording volume control was set incorrectly, or the microphone position and acoustics were wrong.

Amplifier

An amplifier is an electronic device for increasing the strength of a signal, consisting of a valve or transistor circuit. Connect an amplifier between a microphone and a loudspeaker and turn the volume up, and a soft voice into the mic will produce an amplified roar from the loudspeaker.

Essentially the tape-recorder has two amplifiers, the recording amplifier and the replay amplifier, both fitted with a volume control to adjust the amplification. The replay volume control is simply adjusted so that the sound coming from the loudspeaker is comfortable to your ear, but the adjustment of the recording amplifier volume is far more precise. A soft sound arriving at the microphone will produce a faint signal which will then need to be boosted considerably to be properly recorded. A loud crash will produce a heavy signal which will need far less amplification.

The record volume is adjusted so that the loudest sound that will reach the microphone during the recording swings the *volume indicating meter* just to the start of the red stripe on the meter, usually marked zero. The needle should not swing into the red area. So that the record volume does not have to be adjusted during a recording that includes loud and soft sounds, the microphone should be positioned closer to the fainter sounds and well away from any shouting. Often a sound is recorded at lower level than optimum to show the perspective of distance, and it is this ability to adjust the recorded level for artistic effect that gives the record volume knob an important role in making an artistic recording. If you are just dictating a few notes on to tape, then if your machine has an automatic record switch use it as this will ensure every sound is recorded at maximum level.

On stereo tape-machines as there are two channels of sound being recorded and later played back, there will be two separate recording amplifiers and volume controls and two separate playback amplifiers and controls. Sometimes the volume controls are combined into one gain control and a left channel to right channel control to balance the stereo effect.

Additional amplifiers are often wired to the output of modern tape-machines, and include facilities for switching to the radio or record-player and usually have treble and bass controls. However if your machine has a replay amplifier of 3 or 5 watts, or in the case of stereo twice 3 or 5 watts, this extra amplifier is unnecessary and the machine itself has sufficient power to drive extension loudspeakers or the standard pair of stereo speakers.

Automatic Stop and Reverse

Some tape-machines are fitted with special gadgets to switch the machine off when it has run to the end of the spool of tape, or to reverse the machine and change tracks automatically so as to provide a non-stop supply of music. The automatic stop mechanisms are of three kinds: the tape-lace system which keeps the machine running as long as there is pressure applied to a tape-guide by the tape, the metal contact system which switches the machine off when a metal foil strip spliced into the end of the length of tape runs across the contacts, and there is the light-switch system which stops the machine when the tape no longer runs between a tiny light bulb and a photo-electric cell mounted on the tape-deck.

Automatic reverse is sometimes fitted to four-track machines so that they can be used as background music robots. Each end of the reel of tape is fitted with a one-inch length of metal foil spliced between the actual magnetic tape and the leader tape. When the tape comes to the end of its run this metal foil runs through the tape guides completing the electric circuit between two contacts mounted on the tape-deck. The circuit activates a relay switch system that reverses the tape drive and selects a different tape track which has been recorded in the new direction with a continuation of the background music. In this way the machine will play to and fro through the four tape tracks and switch back to track one providing endless canned music.

Autorecord

The autorecord or automatic gain is a device that enables you to record without using or adjusting the normal record volume control. Since the normal control can be set too high or too low resulting in a distorted or hissing recording, the autorecord is ideal for the tape-machine user who simply requires some sound retained and is not interested in any dramatic or artistic colouring of the recording. It is invaluable for recording telephone conversations, meetings and discussions, since it will adjust itself to record the more distant speaker immediately after the speaker closest to the microphone.

Many autorecord switches have two positions, one for use when recording speech, the other for music. The difference is simply in the time taken to increase the volume so that music does not become aggressively boosted the moment a quiet passage is reached. When used for speech, should the speaker hesitate, the autorecord will quickly increase amplification so that background noise such as passing cars will be recorded. As soon as the speaker starts again the unit rapidly brings down the gain but not fast enough to stop the front of the first word being distorted. This results in an edgy-sounding recording which strangely sounds very attractive for sports commentary! The autorecord should be used when it is not possible to predict the strength of sounds reaching the microphone or know where to set the normal record volume control.

Azimuth

The magnetic heads should all stand directly aligned to the tape so that the tiny coil gap down the face of the head is exactly at right angles to the line of the moving tape. If either record or replay head is at an angle to the upright the reproduced sound will have lost all high frequencies and seem woolly, unless of course they are both at the same angle off centre. This can be the case with many cheaper machines that combine the record and playback head. If this head is wrongly adjusted, recordings made and played back on the same machine will sound perfectly normal, but if the recording is played back on a correctly adjusted machine it will sound woolly. Similarly a recording made elsewhere will not replay properly on a misaligned tape-machine.

There are special tapes available with a high frequency tone recorded to use in adjusting the head azimuth. The tape is played on the machine and the small screw at the side of the replay head is carefully turned until the tone can be heard loudest and clearest. If there is a separate record head this can be now aligned by recording the high frequency tone fed in by another tape-machine, and adjusting the record head screw until maximum output is obtained.

Bass Bass is the term given to the lower end of the audible spectrum, and sound-waves on these frequencies require a considerable amount of energy. Once set in motion these waves are difficult to control and will pass through acoustic treatment, walls and buildings in all directions. Listen to the loud record-player next door and you'll notice it is only the low notes you can actually hear.

Bias Each tape-machine contains a bias oscillator which is an electronic circuit generating a signal of frequency far above the audio range. This bias signal is combined with the audio signal before reaching the record head and eliminates low frequency distortion produced in the tape-magnetisation process.

Most domestic machines have had their bias control set at the factory, and it is a difficult matter to find the component to adjust in amongst the electronics. But different types of tape require different settings to the bias control and if you possess a more sophisticated recorder the bias should be corrected to suit the tape being used.

Feed tone into the machine either from a tone-test tape on another machine or from an audio generator. Adjust the record gain volume so that the meter reads —3 and set the machine into recording. Reduce the amount of bias and the meter will fall off to —10. Now increase the bias until the meter reads as high as it will go, and keep increasing until the needle falls back by one division. The bias is now set for that particular tape used during the adjustment.

Birdsong Recording engineers invariably suggest that birdsong should be captured by using highly directional microphones such as the tubular gun microphone, or the paraboloid reflector. Both are expensive, bulky, work very well but tend to be spotted and avoided by the birds. Bird recording is really just an extension of bird-watching. If you know the place and the time that a particular bird performs it is not too difficult to bug the branch of the tree long before the show goes on.

A cassette recorder can be fitted with a long-play tape

cassette which will last one hour. The machine will nestle easily in the tree with the microphone possibly pointing at the singing bird only a couple of feet away at sometime during its hour-long run. Later the cassette is recovered from its perch and played through, any desired birdsong being copied to a master tape reel, the cassette being erased and re-recorded during the next location work.

To obtain good birdsong recordings is a difficult task requiring patience and a knowledge of the birds' habits. Fine recordings are possible using a cassette recorder but as with using far more expensive eavesdropping systems, so much depends upon the birds.

It is often helpful to halve the tape speed when trying to identify a particular bird from its song, particularly to differentiate birds of the same species. The bird will then be singing at half pitch but it will be much easier to pick the song notes apart.

Capstan

The capstan is the rotating spindle from the drive motor that together with the pinch-wheel pulls the tape across the magnetic heads at constant speed. Should dirt or splicing tape adhere to the capstan this will change the speed at which the tape travels, so regular cleaning with methylated spirit is desirable. Oil should never be applied to the capstan.

Cartridge

The cartridge system is mainly used for replaying pre-recorded stereo music in cars. The cartridge itself measures 5⅜″ x 4″ x ⅞″ and contains a continuous loop of tape wound around a central hub, the tape sliding out of the middle of the wind, pulled past the magnetic heads, then wound on the outside of the winding. The system uses special lubricated ¼″ tape that slides easily from the centre of the hub but also leaves its lubrication on the heads which must be cleaned using a special cleaning cartridge.

The eight-track cartridge is recorded in stereo at 3¾″ per second using tracks 1 and 5, 2 and 6, 3 and 7, and 4 with 8. There is a metal foil strip fitted in the tape at a suitable point in the recording which triggers a switch to change from stereo tracks 1 and 5 to tracks 2 and 6, and so on. In this way the eight-track cartridge can provide up to forty minutes playback. Most machines are fitted with a selector to choose which stereo track is to be played as well as the automatic change.

Cartridge recording machines are rare, most users buy pre-recorded cartridges supplied by the gramophone record companies. But there is a similar machine used by radio and TV studios for inserting adverts and jingles into a programme. The tape loop usually only lasts for half a minute, and the recording is preceded by a very high frequency blip. The machine is started, the advert plays and ends, and the blip causes the machine to stop. The cartridge is now removed from the machine with the advert set to be replayed at a moment's notice.

Cassette

The cassette tape-recorder has replaced the standard machine for most home uses. It is cheaper and easier to use, portable, produces good recordings and the cassettes are neat and easy to keep. Providing that you do not wish to edit your recordings, which is practically impossible with this system, a cassette machine will serve the domestic user better than a reel-to-reel recorder.

The plastic cassette contains two bobbins, one of which is wound with ⅛" tape, the end leading past an opening in the box to the second bobbin. The cassette is simply placed over the drive shafts in the recorder which automatically engages the bobbins and feeds the tape past the heads and capstan. In some machines the cassette is placed in the lid which when closed fits the cassette on to the drive shafts. The standard speed is 1⅞" per second.

As with reel-to-reel recorders, the cassette machines are available with autorecord, in stereo, and with automatic stop and reverse mechanisms. But the standard record and replay cassette machine when fitted with a decent sized loudspeaker for playback, and preferably with a better microphone than the ones normally supplied with the machine, is an excellent instrument. See page 55.

Chromium Dioxide Tape

See page 108.

Copying Tape-recordings

To copy one tape to another two tape-machines are required, and an inter-connecting cable. The machine that is to play the original recording will have a socket marked 'output', 'monitor' or 'extension loudspeaker'. An appropriate plug to fit this socket will have to be bought from a Hi-Fi shop together with a cable and plug to fit the 'input' or 'radio' socket on the other machine. The two plugs required may well be of different sizes and types so it is advisable to take the equipment to the shop with you to make sure the plugs and sockets fit.

Having connected the two recorders, set the replay volume on the machine with the original recording to about halfway.If this machine has any tone controls adjust them so that the recording sounds clear and attractive to the ear. Now the tape can be played and the record volume control on the second machine adjusted so that the needle of its VU meter just swings up to the red part of the dial without moving into it. When this has been set, rewind the recorded tape back to its beginning.

The second machine can now be fitted and laced with a full reel of tape after checking that the tape heads and guides are free from dirt. The machine can be set to record and when it is safely running the playback machine can be started. Providing that the original recording is of good quality and the recording machine's VU meter constantly swings healthily towards the red part of the dial then the second tape will be a fair copy of the first. If you are making a copy of a tape that has been made at 3¾" per second, and both your machines are capable of being switched to 7½" per second, then it is possible in half the time. By playing the tape at twice its normal speed and recording it at 7½" per second, when the copy is replayed at 3¾" per second it will sound normal. The only difference that will be noticed from copying at the normal speed is that there is some loss of high frequencies. This can be compensated for by turning up the treble control on the replay machine, or even by turning down the bass control.

Decibel The decibel is the unit for measuring one level of power in relation to another level of power. It can be used for relating any form of energy such as heat or light, but it is particularly useful for measuring relative powers in acoustics and electronics. As the decibel is merely a relative measure, there must be some reference point, but this is an arbitrary point chosen to suit the special circumstances. When measuring sound, the convenient reference point is the threshold of hearing, that is, the acoustic power at which the human ear can just detect a sound. All powers of sound up to and beyond the threshold of pain for the ear can be measured and expressed in decibels with reference to this standard. Similarly inaudible sounds that, of course, are below the threshold of hearing, are below the standard reference point. The difference for audible and inaudible sounds is that those above the threshold of hearing are expressed in a positive number of decibels, and those below as a negative number.

This seemingly complicated measurement of sound is convenient because the ear reacts to sound-waves in a comparative way, and it does not interpret twice the sound power received as twice the loudness. But the ear is also more sensitive to those frequencies between 2,000 Hertz and 6,000 Hertz, so to make sure of a constant loudness a decibel indicator must be used in conjunction with the ear. The measurement in decibels between two sounds will be a comparison of their sound intensity; that is, their power per unit area of the conducting medium, usually air. Loudness is expanded under the heading **Volume.**

With tape-recorders a new convenient reference point is chosen called the zero level. This is the power of signal required to magnetise the tape properly without distortion, and without so heavily recording that *print-through* becomes excessive. Now all levels feeding the tape-recorder are expressed as a negative number, since positive number levels will distort the recording. In comparison with zero level the output of the microphone will be found to be minus 72 dB, which means there is 72 dB gain between the mic and the record head.

It is important not to confuse decibel measurement of relative sound intensities with the decibel measurement of comparative signal levels in the tape-recorder system. The fixed points for each have no relationship, and one is involved with acoustics, the other with electronics.

Disc to Tape Copying

To copy discs to tape an inter-connecting lead must be bought fitted with the appropriate type of plugs to fit the 'output' socket of the record-player, and the 'input', 'auxiliary' or 'radio' sockets of the tape-recorder. If both record-player and tape-machine are stereo, there will be two leads, one for the left or first channel, another for the right or second channel.

With the record-player's tone controls set in their normal position, and its volume set about halfway, play a disc and adjust the tape-recorder's record volume control so that its VU meter needle swings up to the red area of the dial at the loudest point of the music. The tape-recorder can now be switched to record and the disc copied satisfactorily to tape.

If a music tape is being recorded then a system for copying one disc after another needs to be devised to avoid pauses and clicks between each tune. One method is to edit the music together by splicing. Some machines can instantly switch from replay to record without stopping, and if you have one of these there is an easier way than splicing to butt the music together. Fit a felt mat a little larger than the records on to your turntable. You will now be able to start your record-player, but by gently grasping the edge of the felt mat prevent the record from playing just before the stylus reaches the start of the tune. With your other hand play the tape already recorded, and as the music ends, switch the machine into record and simultaneously release the felt mat. In this way one disc can be copied to tape neatly after another, although it takes a few trial runs to master the technique.

Distortion

The two most common types of tape-recording

distortion, and ruination, are caused by simply overloading parts of the tape-recorder's recording apparatus. Firstly, shouting into the microphone or putting it too close to a very loud noise will overload the input stage of the recorder. Secondly, by having the record gain volume set too high the signal can overload the tape-magnetisation process. If the microphone is positioned correctly, and the record volume control adjusted so that the VU meter needle does not swing into the red area of the dial, a good distortion-free recording should result.

Dirt can be the cause of a bad recording. The tape heads soon become covered in a layer of oxide from the tape, and should be cleaned frequently using a cotton bud stick dipped in methylated spirit. If they are not cleaned, soon all high frequencies will disappear from your recording, producing a lifeless woolly sound. Dirt on the capstan and tape guides will cause wow and flutter, sounding as variations in volume and slurred notes.

Microphones are easily damaged, and a faulty mic cannot give a distortionless recording. Many domestic tape users try to use and reuse the same tape year after year which gives them tapes containing no high frequencies and plenty of dropouts. Distortion can occur if the bias setting is wildly wrong for the type of tape used, and many domestic machines cannot be adjusted to suit the new low-noise tapes.

Dolby System See page 109.

Dropout A tape dropout is a momentary break in the recorded signal often caused by dirt on the head or on the tape, a buckle in the tape, or a bad edit. Not all tape-machines will operate with double-play or long-play tape, and will gradually stretch and buckle the tape causing distortion and dropouts. Dropouts can be caused by bad contact between the tape and the heads. This in turn is usually caused by dirt on the tape guides, capstan, or around the pressure pads.

Drama Tapes There are two categories of drama tapes, those that are recorded sound drama not requiring any visual addition to be appreciated, and those tapes that are recordings of actors performing on stage so that they may play back their lines for self-criticism. A stage play recorded is not complete without its visual side and would fail as a radio play.

A story can be adapted for recording by closing the eyes and thinking just in terms of sound. Cut the proceedings into clear scenes that have a definite atmosphere, the first scene in a railway carriage, the second in a church, the third in the churchyard, and so on. This will enable you to simulate the acoustics and background noises of each scene which will stimulate visual images of the different locations in the listener's mind. Keep the number of characters in each scene down to two or three so that the listener does not become confused with the voices, and make sure that your script makes it abundantly clear which character owns each voice. Avoid any involved sequences or descriptive explanatory scenes by writing a narration which will lightly link the drama and help the transition from one imaginary location to another. Keep each scene shorter than four minutes, and rehearse your actors well. Make sure you have chosen a story that includes characters that your actors will know well. If one friend has been to sea let him be the one that plays the part of the boatswain.

A convincing sound drama can be recorded with a reel-to-reel tape machine with microphone and a record-player wired to the recorder as for disc to tape copying. It is all the more convincing in stereo but it is not essential. Each scene is recorded in the appropriate acoustic, and any background effects can be blended in from the record-player, using its volume control to keep the effects lower than the voices. BBC Radio Enterprises produce a series of three special records which contain hundreds of effects. But most effects will be made by your actors. If you want to record a conversation taking place while driving across town, take a cassette recorder and record it as your actors do just that. Finish each scene either with an appropriate sound effect e.g. a door slam, or by fading on speech that contains no real meat so that the listener has dismissed it even before it has faded out.

Let the narration be recorded in a softly furnished lounge by a rich voice that sounds like a story-teller, and who does not appear in the drama. Each scene should be recorded in its proper acoustic. The garden scene can be recorded in the garden, and the church scene recorded in the bathroom which with a little organ music from a disc is quite convincing particularly if your actors speak in hushed tones. Unless you have a properly treated studio it is difficult to simulate an outdoor acoustic indoors. Playing an effects disc of traffic noise behind speech recorded indoors sounds wrong although the listener may not be able to pinpoint what is wrong about it. Even in BBC radio plays the listener can sometimes be jolted back into reality by the use of an improper acoustic.

When recording each scene the producer must put himself in the place of the listener who will not be able to see the expressions, hand gestures or movements of your actors. The only movement that will be audible is that of an actor approaching or withdrawing from the microphone whilst talking. The appropriate sound effects can either be made by the actors themselves, e.g. picking up and answering the telephone, or can be made by an assistant when it becomes more complicated, e.g. walking on the spot producing 'footsteps'.

Each scene should be recorded twice and the most convincing parts chosen from each and edited together. The complete drama can now be edited together with the narration linking each scene. It is important to leave some silence between the drama and the following narration to indicate to the listener that the scene has finished. The final play should be played back a day or so after it has been finished when any meaningless lines or needless narration can be cut out. A short sound drama that succeeds in telling the story and creating the correct atmosphere is far more attractive than the recorded marathon.

Dynamic Range The dynamic range of an audio performance is the ratio between the loudest and the softest sound produced, and measured in decibels. The dynamic range of an orchestra can be 70 dB between the lightest passage and a full crescendo. Such a wide dynamic range is difficult to record, because if the record volume control is set to properly receive the loudest passage then the solo flute three minutes later will be recorded at such a low level as to become lost in the background noise of the recorder.

Modern recorders have a *signal-to-noise ratio* of around 50 dB, which means if the softest sound recorded is 50 dB or more down on the loudest sound then it will be lost with the tape hiss. The answer to recording music, drama, speeches, or sound effects that have a wide dynamic range is to carefully adjust the record volume control to amplify the quieter sounds being careful not to over-record the clamour. Of course there should still be difference in the volume between loud and soft, but you will be able by amplification of the weaker sounds to compress the recording to a dynamic range of around 20 dB, which will replay satisfactorily. With music it is necessary to know when the volume will suddenly increase so that the control can be slowly turned down and not produce a jerky recording. The human ear cannot register differences of 2 dB or less, so the control should ideally be moved up or down in steady 2 dB stages.

Echo An echo is a sound-wave reflected to the listener and arriving slightly later than the direct sound. If the reflecting surface is quite distant, such as in the case of echoing rock formations, the time for the reflected wave to arrive will be quite long, and a complete repetition of the original sound will be heard. Nearer reflecting surfaces will produce a reverberation effect that colours the original sound since the reflected wave is only arriving slightly behind it. The effect is quite easily produced by singing in the bath, but for recording purposes it needs to be controlled.

Some tape-machines have a facility to blend the output from the replay head with the signal being recorded. As the replay signal will be slightly behind the recorded signal this produces an echo effect known in recording studios as *tape-echo* or *flutter-echo*. The effect can be varied by increasing the amount of feedback or by varying the tape-speed, and the actual sound obtained will depend on how far the record head and replay head are spaced apart on any particular machine.

Professional recording systems use other forms of echo to colour the singing voice, often taking more than one effect on the same tape. Spring-echo is produced by applying the original signal to an electro-magnet which causes a spiral spring to vibrate. At a further point on the spring is a pick-up which is wired to feed the delayed and vibrating signal back with the original. Plate-echo gives a different effect but works in a similar manner, in this case the spring being replaced by a thin metal plate. A hollow sounding echo can be produced by using an echo-chamber. The chamber can be an empty room, cellar, or a metal tank insulated on the outside to keep the echo in and extraneous noise out. The input to the chamber is a feed of the sound that is being recorded, let us say a singing voice. The feed is applied to a loudspeaker at one end of the chamber and the voice reverberates around the enclosure to the far end where a microphone is fitted. The signal from this microphone can then be mixed in the desired amount with the original signal direct from the singer's microphone. Each echo-chamber will colour the sound in a different way depending on its size, shape, and how the microphone and loudspeaker are mounted inside it.

Echo can easily be added to any recording by one of these methods, but if there is reverberation already present from the hall or room in which the performance is being given, this cannot be removed. For this reason studios employ a fairly dead acoustic and add the finely judged amount of echo electronically.

Editing Editing will not cure a bad recording, it cannot unmake distortion, remove unwanted echo, or eliminate tape hiss. But editing can make a tape far more attractive by removing the pauses, the repetitive phrases, and unwanted clicks. Digressions and mistakes can be cut out, and sound effects and music cut into the recording. It is possible to speed up a pedantic speaker by removing the gaps between words, and to make a stammerer intelligible by cutting the unwanted sounds. During a production, if a mistake is made the speaker can pause and repeat his line correctly and the fluffed line can be removed later by cutting. It is the skill of editing together with artistic mixing of sounds that really separates the professional sound producer from the amateur.

Editing is not difficult to learn, but it needs practice so that the ear can become used to operating in conjunction with the hands. The tools required are a yellow chinagraph pencil; a reel of adhesive splicing tape; a single-sided razor-blade; and a splicing block. There are many types of splicing block on the market, but the one I recommend, and which is used exclusively throughout the BBC, is the EMI designed block. This has a channel its complete length to grip the tape in position, and a cutting slot at 45° to the tape for slicing the tape with the razor at an angle.

The recording is played through and points in the speech where the cuts should be are noted. For example, if the speaker is forever repeating the phrase 'yer know' through the recording, then it will be decided to cut out every second one. The points to cut the tape will be immediately before the y of 'yer', and immediately before the next word that follows 'know'. If you were to make the second cut bang after the word 'know' then the edited tape would contain two breath pauses, one each side of the join. Having decided on the points in the speech where you will cut, play the tape until the first point is reached. Press the pause button on the machine, and move the tape gently to and fro by slightly rotating the tape reels with your hands. In the loudspeaker you will be able to hear the point at which the word starts and move the tape very slightly so that it is situated just back from the replay head.

With the cutting point now immediately above the replay head, it can be easily marked by drawing a line across the tape sitting on the head. The tape is now played on until the second cutting point is reached and this is marked in the same way as the first. Often the two cutting points will be so close together that the tape deck should be left in the 'pause' position, and the tape moved on by hand.

Now that the cutting points have been marked, the machine can be switched off, and the tape unthreaded from the heads. It is placed in the splicing block so that the chinagraph mark is above the cutting slot, and is sliced at 45° to the vertical with the razor-blade. When the two cuts have been made the independent piece of tape that has been removed should be taken from the block and kept. The two free ends of the edit can now be spliced together by pushing them close together in the block. There should be no gap between the ends, nor should they overlap. The ends are stuck in place by pressing down a ½″ strip of adhesive splicing tape over the join. This sticky tape should sit neatly over the magnetic tape and not protrude in any way. Of course the splice is made with the rough oxide side of the tape facing downwards in the block, so that the splicing tape has been stuck over the shiny backing side of the tape. Splicing tape should never be stuck on the oxide side as it will obscure the tape from the heads. The edited tape should now be played back, and if there has been a marking error then the removed portion of tape can be spliced back into the whole, and the marking process repeated until correct.

Once you have started editing you will begin to recognise the sounds of individual word letters, even when they are played at low speed or backwards. Gradually the initial difficulty that everyone has in finding the cutting points will pass, and you will be able to cut tiny fractions of tape, such as the 's' off the end of a word.

A tape may sound obviously cut about because the editor has not left the correct amount of pause between words, or allowed the speaker a proper breath between sentences. If a recording made in one acoustic is joined to another made in another acoustic, then the join will be obvious. If two separate tapes are to be spliced together then they should be at the same volume or there will be a change of level at the joint.

Editing can only be performed on a tape that has one track recorded its full length. If your tape recorder is half track (i.e. twin-track) then the bottom track must be left unrecorded as the cutting would destroy it.

It is impossible to edit cassette tapes without very specialised equipment, and it is usually unsuccessful because of the narrow tape and low speed of the cassette-recorder. Many home machines have enclosed head units with pressure pads that make marking the tape on the replay head impractical. If you have such a machine then off-set marking must be employed. This consists of marking the tape at some fixed point on the tape deck which is quite close to the replay head, for example at the capstan. Now the distance between the replay head and the capstan must be accurately measured. If a mark is made on the editing block the equivalent distance to the right of the cutting slot, the tape mark can be placed at the point and the edit will be made in the correct spot.

If you have an important recording that is full of pauses, ums and aaahs, then possibly the best method to clean it up is by dub editing. This necessitates having two machines, one of which can be flicked from playback to record at the press of a switch without stopping, and without recording a click on the tape. Many of the more expensive domestic machines on the market favoured by stereo enthusiasts are of this variety. The technique is to copy the original tape to the second machine. When there is a pause or fluff in the speech both machines are stopped. The play-in machine is set in the pause position immediately after the offending material. The recording machine is spooled back and set to play. When it reaches the point at which the fluff or pause occurred it is instantly switched into record. At the same time the play in machine is released from the pause switch and starts feeding the other recorder. In this way the copy will finally appear edited of all mistakes, and the original recording can be erased and the tape reused.

Editing is used not only to make a recording word perfect and to bring separate recordings together, but also to make the tape more listenable. If you play a tape through, it may be correct in every detail, well recorded and without repetition or fluffs. It may also be boring simply because it is too long. If this is the case then sections of the tape must be sacrificed, and these must be chosen very carefully so that they do not detract from the meaning of the whole. Often a very interesting anecdote or digression must be cut even though it sounds so good, simply because its inclusion will make the length of the whole piece too long.

When editing non-scripted material it is important to remember that the occasional hesitation and pause is part of the speaker's character. If every single natural mistake is cut from the tape then the person that was speaking into the microphone will eventually sound like a robot. However, during scripted recordings most pausings will be caused by the reader being unsure of his script, and it is best that he goes back to the beginning of the sentence in which he fluffed, and the mistake edited out later. Music editing is a very rewarding task, and with a little effort quite amazing non-stop party tapes can be created although you may have no facility for mixing two record players. The skill is to copy a disc to the tape that starts in the same key that the previous disc had ended in, and then to cut the two together so that the two tunes keep in beat.

Effects

Any recording that is intended to communicate with a later listener will improve with the addition of appropriate illustrating sound effects. Correctly used, a sound effect can animate a recorded talk holding the listener's attention. Their use in sound drama enables the producer to convey a certain situation or location instantly and without narration.

There are sound effects discs on the market which contain many useful recordings including exciting illustrations such as a lion's roar which would be difficult to obtain oneself and impossible to simulate. But there are those sounds that are best produced live at the recording and sound faked if mixed in from disc. These include those sounds that are meant to be produced by the actor or speaker involved, when he is unbolting the door or answering the telephone for example. On these occasions the microphone should be so placed that he can produce these noises for himself using a real bolt on a door, or telephone. Footsteps are difficult to simulate, and usually those recorded on disc are in a different acoustic to the actor's voice and are therefore unsuitable. A reasonable effect can be recorded by placing a second microphone near to a stone slab or wooden plank, or even tray of gravel, on which your assistant walks on the spot. He can be directed to start, stop, or quicken his pace by the actor reading his lines at the other microphone. If it is impossible to arrange for two microphones to be used then the effects assistant must perform his footsteps near to the actor's mic, and this can be accomplished by elevating him onto a table top.

Many noises do not appear to be what they are when replayed from tape. A rifle shot does not appear nearly so dramatic as it should, and a good sound to replace it is the bang from a toy percussion-cap pistol. A roll of thunder can be made by rolling beads or dried peas across a tambourine, and distant gunfire by dropping the peas onto the skin with the tambourine held close to the microphone. An arrow flying through the air is faked by swishing a thin cane or stick past the microphone, and as it reaches the tree at which the archer has aimed the effects man can firmly stab at a book cover with the end of the cane, producing the required quick dull thud. The guillotine scene can be simulated by sliding a carving knife down its sharpener, finally chopping it into a potato to complete the decapitation sound. Walking in the snow is beautifully reproduced by holding a long piece of cotton wool near to the mic and twisting it to and fro. Cellophane paper being screwed up close to the microphone sounds like a fierce forest or house fire. And everyone knows that two halves of a coconut shell tapped together becomes horses' hooves, and a matchbox half-full and shaken up and down becomes a marching army.

Long background effects, such as rainfall or a travelling train, are best played in from a disc, but if the particular location effect can not be found it is always possible to record the effect on a cassette machine and play it in later from the cassette. If you need a certain location's background noise then the best atmosphere to record is the real thing, and a cassette made in a real harbour will generally sound far better than mixing together sea noises, ships' sirens and seagulls. But it is possible to choose similar substitute locations to record. Any piece of electrical machinery working in an echoey building will sound like a full factory. Early morning birdsong, without the traffic noise, produces the loveliest country garden. And a recording made in any large supermarket can become a hotel bar, an airport lounge and even a supermarket!

Equalization

The magnetisation process in the tape-recorder is by no means perfect, and if a recording signal is directly applied to the recording head, and the replay signal taken directly from the replay head, then the reproduced sound will appear very different from the original. During recording the signal loses its higher frequencies, and during replay it loses its lower frequencies but boosts the signal more as the frequency increases. To off-set these changes in the system's frequency response, the tape-recorder has two frequency discriminating circuits built into the process, one before the recording head and one after the replay head. These circuits affect the signal so that on replay the sound is similar to that recorded, without increase of bass or treble. They are the record equalizer and the replay equalizer, and their use is known as equalization.

Fault Finding

If your machine fails to work, or works badly, check through this list before assuming it is something complicated requiring an electronics engineer.

If the machine will not run or spool and it is a battery recorder, check that the batteries are new and firmly placed into their compartment in the correct manner. If it is a mains machine, check that the wall-switch is on, that the socket will operate another appliance, that the mains fuse, the plug fuse and the recorder's fuse are in place and in order.

If the machine records or replays a very woolly sound lacking treble, check that the tape is laced the correct way around with the shiny side away from the heads. Loss of treble is often caused by dirty heads as home-users forget that the heads should be regularly cleaned with cottonwool buds and methylated spirit. A woolly sound will result from using old tape, or if the tape is stretched or twisted. A drop in high frequency will occur if the bias current has been adjusted too high, or if the replay and record heads are misaligned.

When the machine replays a recorded tape, but does not record, check that the record volume control has been adjusted, and then that the microphone is correctly plugged in and any input switch moved to the mic position. Check that there is no obstruction between the tape and the recording head. If the machine is recording only a low hum then the microphone cable may be at fault.

If the recorder records but will not erase, try cleaning the erase head. The erase circuit will not operate properly if the supply current is low. Usually the erase oscillator valve or transistor is the first to break down in a machine as it is pretty hard-worked, but if your recorder is fairly new loss of erase will probably be due to dirt on the erase head or some obstruction between the tape and the erase head.

Often the user will believe his machine is faulty if he is not fully familiar with using a tape-recorder. The recorder will not operate unless the tape is correctly laced around the tape guides, across the heads, and in between the capstan and pinch-wheel. The record and replay volume controls must be adjusted, and the microphone and loudspeaker connected properly. If it is a battery machine then the batteries must be replaced frequently.

Flutter

Flutter is rapid variations in tape speed, which produces frequency fluctuations at the output giving a warbled effect to the sound. Modern recorders produce a negligible amount of flutter, and any speed distortions will be caused by dirt on the capstan, tapeguides or heads, or perhaps using different sized spools on either side of the machine.

Four-Track

The term 'four-track machine' can mean two things depending on whether one is referring to domestic or professional equipment. In the professional sense it refers to a recorder that has four record/replay heads one above the other so that the machine can record four separate channels simultaneously. This is used in music studios where each track carries a different instrument. When the tape is played back the tracks can be properly combined and balanced to provide an attractive sound long after the musicians have gone home. This type of machine, which should be referred to as a 'four-track, four-channel' recorder is also used for **Quadrophonics** which are covered under their own heading.

The second meaning of 'four-track' is that commonly applied to domestic recorders. These record the length of the tape four times over. Each track takes up a quarter of the tape width, and two are one direction separated by another track recorded in the opposite direction. The fourth track lies at the bottom of the tape in the same direction as this separating track. To accomplish this track pattern, the domestic four-track machine is fitted with two record/replay heads mounted above each other and separated by half the tape width. The machine can be switched from one head to the other as desired, and in doing so changes tracks. The other two tracks come into being when the tape is turned over and played in the opposite direction.

Any slight fault in the tape will be four times as damaging with this four-track system, so it is important to use new tape of high quality.

Film Tapes

A home movie will greatly improve with the added dimension of sound. It is not necessary to purchase an elaborate sound projector to be able to add music, sound effects and narration to a short film. For most amateur films it is not important that the tape recorder and projector are not synchronised providing that the 'sound track' has been produced with this in mind.

The film will normally have a general theme and setting, and the mood music can reflect this. For most of us home movies are a record of holidays abroad, and the music should be chosen to identify the place and indicate that it's a happy time. Let us suppose you intend showing a film of your holiday in Spain. The sound track could start with a lazy Spanish tune that spells out sun, sea and sand. After the musical phase it can be faded down beneath your narration that should indicate who the people are on the film, and where it was shot. This voice piece should be kept fairly brief and the music brought up again.

Sound effects can be slowly mixed with the music to help the atmosphere. If the film starts to concern itself with city sights, then traffic noise can be gradually blended with the music. Similarly if the camera starts to concentrate on the beach, then the sound of lapping waves should be fed in. These effects should be brought in only little by little so that it will not disturb the viewer if the correct scene has not yet appeared. The approximate spot for each sound and new narration can be found by running the film and timing how far from the start each new scene appears. When the tape is made each segment can be timed so that it is reasonably in step with the film. As there is plenty of music between each piece of narration and the effects are just edged in it does not matter that the tape and film are not synchronised. The film script should be to the point and avoid phrases such as 'now you see . . .' in case the picture has just divulged whatever you were about to introduce to your audience. Care should be taken not to state the obvious since the audience can glean a good deal of information for themselves from the screen. With holiday films full of family and sights the narration can entertainingly consist of anecdotes about holiday happenings not necessarily connected with the shot at all.

The film tape should be replayed through a speaker situated at the side of the screen. The technical aspects of producing the tape are explained under the heading **Mixer**, and further useful information is given under **Scriptwriting** and **Uses.**

Frequency Response

The frequency response of your tape-recorder is its ability to record and reproduce the sound spectrum. The recorder's limitations show at the bottom and top of the scale, and its performance will vary according to the tape speed. For home use a frequency response from 50 to 16,000 Hertz, or cycles per second, is quite satisfactory and transistorised equipment will usually provide this performance even with the machine running at 3¾″ per second. The tape machine manufacturers state its response within a certain tolerance of reproduction which should be not greater than plus or minus one decibel. Often the higher end of the response is stated as 21 or 25 kHz, which is quite possible but since most of us have trouble in hearing 14 kHz there seems little point in comparing the performance of different machines over 20 kHz, although the manufacturer with the best frequency response would appear to be making the best machine.

If your recording system is incorrectly adjusted its frequency response will immediately suffer. The high frequencies will disappear if the heads are not aligned, if the bias current is too high, if the tape is old, or there is dirt on the heads.

Full-track

Full-track is the system adopted in radio production, and is particularly useful as the tape can be cut and edited. It consists of recording a track the full width of the tape which although it safeguards against dropouts, means that the tape can not be turned over and played back onto its original spool as with multi-track systems.

If you want to edit your recordings there is no necessity to convert your machine to full-track as all that is required is a single track of recording. This can be obtained by wiping both tracks, or in the case of a four-track machine erasing all four, by switching the recorder to record with the record volume control at zero. Alternately virgin tape should be used. Then with a completely clean tape a single track can be recorded, and can be cut and edited without interfering with an adjacent recording.

Gain

Gain is the amplification given to any particular signal to achieve a good recording. A very soft sound will produce a minute signal in the microphone and if this is to be replayed satisfactorily then the record volume control must be adjusted to a high position to increase the gain. An automatic record device will adjust the gain automatically so that every signal is amplified to a common level.

If a number of different voices are being recorded, for example in a discussion, the gain can be affected by adjusting the position of the microphone. With the record volume set correctly for use by the strongest speaker at three feet from the microphone, the appropriate gain for the softest voice can be obtained by moving the mic to within nine inches of the lighter speaker. This will produce equally strong signals in the microphone so that each voice will be correctly recorded without movement of the record volume control.

Half-track

Half-track or twin-track recording consists of a system whereby the recording heads only magnetise a track along the length of the tape that occupies half the width.

It is the common method employed in home recorders, including cassette machines, as it has the convenience that when the tape has run to the end the full take-up spool can then be turned up-side-down and put onto the feed side of the deck. When laced-up the machine will now read the second track which runs in the opposite direction. This system requires half the amount of tape that would be used in a full-track recorder, but it has the disadvantage that the tape cannot be cut for editing purposes.

If your machine is half-track and you wish to edit the tape after recording, then only one track must be laid. This will leave half of the tape empty of information, but it can be cut and spliced without fear of destroying the recording that would normally be laid on this other half.

In stereo systems half-track recording is effectively four tracks being laid on the width of the tape two at a time, the upper pair in the opposite direction to the lower. The two tracks on the upper half of the tape are the two channels of one recording, and on the lower half of the tape are the tracks of the separate stereo recording laid in the opposite direction when the reel has been turned over. This arrangement means that the stereo tape can be played successfully on a mono recorder, the half-track head reading and combining the stereo tracks.

Heads

The magnetic heads that erase, record and replay the tape are small electro-magnets. An electro-magnet consists of a core of iron on which is wound a coil of wire. When an electric current is passed through the coil, a magnetic field is set up around it, and the core behaves like a magnet. Magnetic heads are designed in a ring shape with a tiny gap in their core across which the magnetic fields they produce have to jump. This gap on the head is positioned so that the tape passes across it, bridging the non-magnetic gap and in the process becoming magnetised itself. In this way the tape-recorder erases and records on magnetic tape.

In the case of the replay head the process is reversed. A magnetised tape (that is a tape that has already been recorded) is passed over the gap of the replay head and in doing so feeds its magnetic field in the coil of the head. When a magnetic field is moved into a coil of wire it induces an electric current in the wire, and it is this induced signal that is amplified, fed to a loudspeaker and reproduces whatever was recorded on the tape.

The magnetic heads are so arranged in the tape-recorder that the tape is rubbed closely over each tape face in turn. The centre of each tape face is the gap. The first head that the tape glides over is the erase head that will wipe any recording on the tape, providing that the machine is switched into the record position.

When switched to record, a circuit inside the machine feeds a current to the erase head which is rapidly changing from one direction to the other. This is known as a high frequency alternating current. This produces in the erase head a rapidly changing magnetic field. As the tape passes across the head gap this alternating field attempts to align and realign the ferric oxide crystals of the tape. These crystals react to a magnetic field lying in one position, when immediately the field changes and they adjust, by which time the field has already changed again. Because of the physical limitations on crystal movement and because the magnetic field is changing so rapidly, the tape leaves the erase head with its ferric oxide crystals lying at random and so completely demagnetised. In this manner any previous recording is wiped from the tape and it is ready for recording.

The tape now rubs across the recording head where the magnetic fields produced in the head by the recording signal pass through the tape and magnetise it by aligning its crystals in a certain direction. The tiny portion of tape that is over the head gap at one time will be magnetised according to the magnetic field produced in the head at that time, and the crystals of this portion of tape will now behave like a magnet with the same field. In this way the tape becomes an endless line of tiny magnets, each one a record of a particular magnetic field that was produced in the recording head when it was against the head gap.

On some machines the tape passes from the recording head to the replay head so that the user can immediately monitor a recording that is being made. However most domestic machines use a combined record and replay head, and the same head will read the tape when the recorder is switched to playback. A combined head cannot do both jobs at the same time and so listening to the quality of a recording while it is being made is impossible.

To improve frequency response modern magnetic heads are being made with a smaller and smaller gap. A point is reached when the gap-width does not allow for an adequate bias field to be produced, and the bias current cannot be increased further as this causes loss of treble frequencies. (See **Bias**.) To enable the head gap to be made really narrow some more sophisticated machines have an extra head situated almost opposite the record head. This *cross-field head* operates from the other side of the tape to normal, and is mounted at an angle away from the tape direction. By feeding this head with bias signal the bias field of the recording head can be extended. This enables a narrow record head gap to be used and a better frequency response obtained, which is particularly useful at low tape speeds.

After continual use the recorder's heads may become permanently magnetised to some degree. When this happens each head will not only have a magnetic field produced in it by a signal, or the passing of the tape, but will have a field of its own. This will add a hiss to any recording, and will gradually erase tapes that are simply being replayed. There are several inexpensive demagnetisers on the market which are easily operated and run from the mains. When switched on the poles of the demagnetiser should be brought slowly up to the head and slowly withdrawn to effect the cure. It is advisable to remove your watch during this operation as even non-magnetic watches can be damaged by exposure to intense magnetic fields. (See **Azimuth**.)

Hiss

Tape hiss is the most prevalent background noise of the

tape-recording process. If your recording has been made correctly, the sound will be so much higher in volume than the background noise that the latter will be unnoticeable. However, if the recording has been made with the record volume control set too low, the bias incorrectly adjusted or with the magnetic heads misaligned, then on replay the sound will be at a low level. When this is amplified to a listenable volume the background hiss will be amplified at the same time and will be clearly audible, marring the sound.

Old tape or recorded tape that has been stored will appear hissy. If an old recording is particularly valuable it can be replayed through a top cut filter which will remove all the higher frequencies including the hiss. If your machine has a replay tone control this can be adjusted to cut the treble frequencies so that the hiss is less apparent.

Hiss will appear if one or more of the heads is magnetised, which is discussed under the heading **Heads.**

Hum

Hum is commonly caused by a mains cable being adjacent to the low-level parts of the recording apparatus such as the microphone, the microphone cable or any leads connecting radio or record-player to the machine. Any minute currents induced into these low-level areas by a power cable will be amplified considerably by the recorder and become audible as hum. The magnetic heads should ideally be housed in a screening metal cover to prevent any interfering induction occurring in the replay head circuit. If your machine has only a plastic cover over the heads, then it is unwise to bring this area close to any mains cable or, for example, operate the machine near a table lamp.

If the outer sheath of the microphone or radio cable is damaged then the recording will be at a very low level and covered by a strong hum. Often the hum is being injected into the system by the recorder's own mains cable and this may be cured by reversing the position of the two wires inside the mains plug.

Hum can be induced into the recording system by

operating your recorder close to another piece of apparatus. Often the machine is placed on a shelf alongside the television, radio and record-player which is the worst position for picking up hum and other interference. Electric heaters and fires, telephones, refrigerators, and even an electric train-set will induce hum into any audio system if in close proximity.

It is possible to record and find a low rumble beneath the replayed sound. This can be caused by dirt on the capstan or pinch-wheel, or by positioning the microphone too close to the recorder where it will pick up the noise from the motor. Rumble is frequently heard on tapes that have been recorded from discs. This can be traced to the record-player's motor, and the cure is to earth the turntable by fitting a wire trailing in contact with the inside of the turntable while anchored at the other end to the unit's chassis. If it still persists, then try reversing the wires inside the mains plug of the record-player.

Stereo systems seem to be beset by hum which is not surprising since there are usually so many separate units and jungles of inter-connecting leads and sockets. The fault is commonly that one of the leads has a break in the outer earthing sheath of the wire, or that one of the plugs has become disconnected from the wire sheath. To clear the fault try making up a new connector and replacing each original connector one by one until the hum disappears.

Interference

Even though your recording apparatus is screened effectively from any source of hum, it may suffer from radio interference particularly if you live in a town or city. Any close radio transmitter such as a taxi, police or ambulance radio-telephone will be at sufficient power to be picked up by the electronics in your recorder and fed on to the tape. Television sets in operation transmit a signal which interferes with tape-recording as a high-pitched whine or tone. If you can only record near these sources of interference then the recorder should be completely screened. This can be arranged by fitting the machine into a metal box which is then connected to the third earthing pin of a mains socket. Aluminium foil can be used to line the plastic case of a recorder and, providing that it is earthed, will screen the electronics from any radio interference.

Interviewing

A sound interview is an excellent way to convey information, but the production of a good interview is a far more difficult task than one first imagines. The tape-recording needs to be clear and an acoustic background that lends itself to the interview. Grandma will sound enthralling chatting in her country garden, but recorded in a dead studio she may seem ponderous or lecturing. It is natural for a racing driver to answer questions as he drives along with all the engine noises and gear changes adding to the recording. But these same background sounds would be ruinous to an interview about ballet. The choice of correct acoustic is an easy matter for the interviewer with a portable recorder, as the interviewee will normally be at or near his normal place of work and in a place with the appropriate acoustic for the topic to be discussed. A scientist sounds more convincing in his echoey laboratory, and the pop-star becomes loaded with glamour when recorded with all the background sounds of a night-club. The interviewer must insist on recording the tape in the interviewee's real world, or at least his projected world, rather than being taken to an office with the wrong background and acoustics. BBC practice for many years was to record location interviews in the nearest softly furnished room, the argument being that a clear recording totally lacking in colour was better than an interview badly recorded or containing obtrusive background sound. It is now generally accepted that by holding the microphone near to the speakers, and by standing a reasonable distance from anything really noisy, a more animated and interesting tape can be produced.

In quiet surroundings the microphone can be held equidistant between interviewer and interviewee, or slightly more towards the one with the quieter voice.

With high background noise, the microphone should be moved smoothly with one arm movement from speaker to speaker. If the microphone is moved suddenly this will probably result in clicks being heard on the recording, so a practised microphone arm can save a lot of later editing. The mic should be held firmly and as still as possible for each speaker to keep the volume constant. The microphone cable can be looped around the hand once to prevent it from swaying or pulling at the connector on the mic. The microphone itself should be placed slightly towards one side of the speaker's mouth, not pointing towards it. This helps avoid blasting when certain words are used that cause the mouth to expel air forcefully, such as 'pop' and 'pepper'. Also having the microphone to one side the interviewee is less aware of it than if it is placed directly between your two faces.

As the microphone is being placed so close to the source of sound, the record volume control will be set lower than normal. Providing that it is set correctly a good recording will result with the background noise at an acceptable level even at a rowdy party or on an active building site. Close mic technique also enables the interviewer to conduct the proceedings by moving the microphone. This can be particularly useful with an interviewee who wants to go into fine detail about every small point, or wants to develop his answer into a long speech. When he sees the mic being moved away as he digresses he will soon come to the point.

When you are in the correct surroundings for the interview, conduct a short pre-recording test by taping the first question and answer and playing them back. This will help relax your interviewee by involving him in the technical side of the proceedings. More importantly it will enable you to check the recording, listening for any distortion, which voice was louder, and if the voices were clearly audible and understandable above any background noise. You will also be able to hear how the acoustic really sounds, as sometimes a particular location sounds like something else, for example an aeroplane hangar close to the whine of jet engines sounds more like a wood-mill!

If there was distortion during the test, turn the record volume control down a little. If one voice was more outstanding than the other, make sure that the microphone is brought closer to the weaker voice on the re-recording. If the background noise swallows the interview, you will need to position the microphone closer to the speaker's mouth, and necessarily turn the record volume control down. It is impossible to record a proper interview and be gazing at the VU meter throughout, so this pre-recording test is essential.

If the level suddenly increases during the interview because your speaker suddenly becomes excited or dramatic, the microphone should be smoothly drawn away from him to compensate. Similarly if Grandma becomes hoarse, then the mic should be brought closer to her mouth. The pre-recording test will also show if the tape has been incorrectly laced and if there is any interference. Before an interview at a London embassy I found the machine was recording Morse code messages being sent across the world from an adjacent wireless room.

The interview itself needs to be interesting and communicate a certain argument or information. This can only be done if the interviewer conveys to the interviewee a genuine desire to know the answers to his questions. So the preparation for an interview should not consist of researching and learning the subject, but deciding what you and your audience really want to know about the topic and making sure your interviewee satisfies the questions. The interviewer should listen very carefully during the recording and become wholly engrossed with the subject, and not be reading his question notes or worrying about the VU meter.

During an interview the information should come from the man that knows, not the interviewer. The interviewer's job is to spur the speaker into enthusiastic explanation and guide the proceedings so that all sides of the issue are covered. The interviewee will be able to relax and speak effectively if he is convinced that you are listening to him. Look him in the eyes, listen to what he is saying and play the Devil's advocate. Ask him why things are so, how, when, where and who, and express any natural objections that may arise from anything he might say. Do not ask long questions that display how well you know the topic as the answer may just consist of 'yes' or 'no'. We have all heard the young radio reporter welcoming a returning ocean yachtsman with: 'John Hatch, welcome back to Britain after your three-month cross-Atlantic voyage in Esmaralda your thirty-six foot catamaran. Is it true you've had force ten gales and trouble with icebergs?' The answer comes back: 'Yes'. An interviewee should never be allowed to read from a prepared script, or be allowed to hold the microphone. Writing for speech is a specialised art, and inevitably your speaker will sound as though he is reading. This will result in loss of enthusiasm and humanity, and the subject will not be conveyed to your listeners even though your speaker's notes are written in beautiful English.

If the interviewee takes hold of the microphone you can explain that it is necessary for you to hold it to balance the sound. Failure to do so will probably result in the speaker holding it too close to his mouth, or waving it about producing clicks and fading of the speech.

When arranging and meeting the speaker, avoid using the word 'interview', as this conjures up images of interrogation. Explain that you would like to chat about the subject with the speaker, and record the conversation. He will usually want to know the questions you will be putting to him. To emphasise that he is not about to be grilled, tell him that your listeners are interested in knowing about the topic or issue in question, so if you both were to talk around the subject everything would be explained. As a preface to your leading first question in the interview, describe your speaker with his full title, post or accomplishments as this will build up his confidence and produce a better tape. Do not allow him to sit behind his desk but bring him on to common ground in an adjacent chair. If the interview will sound better in an alive acoustic that adds to the subject, move into it. If the recording has to be made in an office, make sure that you are not interrupted by telephones or secretaries. Politely request that your interviewee does not smoke his pipe, fiddle with his retractable pen, or rest his head between his hands, as they will all affect the recording. The keyword for a successful interview is enthusiasm.

Jacks

Jacks are the multi-way plugs and sockets that connect each piece of apparatus to their inter-connecting cables. In this way the microphone, radio, record-player, and any external amplifiers and loudspeakers can be securely brought into the electronic circuits of the tape-machine. There are many different types of jack in use, ranging from the professional Post Office jack to minute phono-plugs. This means that inter-connecting different makes of equipment can be quite difficult as they may well have different types of jack socket fitted. Some of the larger Hi-Fi shops keep a large stock of inter-connecting cables fitted with various combinations of jack plug but it is as well to find out the size and name of the sockets on your equipment before going along to buy the jacks.

If there is a bad connection between the connecting cable earth sheath and its jack plug, or if there is dirt between the plug and socket, then your sound system will suffer from hum. The small plugs and sockets are best cleaned with methylated spirits on cotton wool, while the larger jacks need polishing with metal cleaner. See page 45.

Keeping Tapes

Tapes are not a permanent form of recording since the magnetic field of any portion of the tape will be slowly remagnetised by the rest of the reel, and itself will be affecting the tape around it. The earth's magnetic field will also be acting on the recording, so that after a number of years the original tracks of magnetisation on the tape will have become partially erased. To extend the life of a tape it should be played through at least every six months to rewind the tape into a different placing. It is imperative that your archive tapes are kept well away from magnetic fields such as those produced by electrical appliances. An audio engineer acquaintance of mine kept his recordings on a bookshelf situated above his stereo loudspeakers. Within a month the magnetic fields from the speakers had erased the complete tape collection.

The worst enemy of tape-recording is dirt, and tapes should be kept in polythene bags and boxed. The outside of the box should carry a list of items recorded on the tape with an indication of how many minutes from the start of the tape each is located. Recordings that you wish to be kept a life-time, such as a wedding ceremony, should be transferred from tape to disc as this is a permanent mechanical form. Most professional music studios can arrange for a single disc to be cut, and a full hour on a 33 r.p.m. record would cost around £12.

Leader

Leader tape is coloured plastic tape similar in width and thickness to magnetic tape, but of course incapable of being recorded on. It can be spliced before and after a recording to ensure complete silence, free from switch clicks. It is particularly useful if there are many short separate items following each other on the same reel. By using different colours the leader will enable you to identify the tape band that is spliced after it.

During a drama production or a radio programme it may be necessary to build up one or more insert tapes. These will contain in the correct order all the sound effects or interviews that are to be played into the production. Between each segment is spliced a foot of coloured leader on which to pause the play-in machine, and to show visually when the insert is about to end.

Letter Tapes

The tape-cassette is the ideal form of audio letter as it is small and cheap to mail, and all cassette machines run at the same tape speed. Ordinary tape can be wound on to a three-inch reel for sending through the post, and this will hold a reasonable length letter if recorded at a slow speed or particularly if it is half-track. It is important to find out whether your tape correspondent has all the tape speeds that your machine will switch to, and if there is any doubt record at 3¾" per second, as this is the most common speed.

Once the recorder is started it will be difficult for you to talk into the microphone constantly unless you have prepared the letter. If you write a script then the tape will tend to sound lifeless, and it is far better to make a running order of the subjects you want to mention, and under each subject a few notes on any details. Then with the machine set to record and yourself nicely relaxed in an armchair with the microphone placed about a foot away, speak as though your tape correspondent was actually present. Develop each subject one by one and put in absolutely anything that comes to mind as though you were in conversation, but in this case your listener cannot answer back! As with a normal letter if you expect a reply do not forget to include your current address, and also a note as to the tape speed.

Limiter

A limiter is an electronic circuit that keeps the levels of a signal down to a prescribed limit. This automatic level control device is particularly useful in that it can prevent a recording system from overloading and ruining the tape with distortion. The limiter is often combined with an automatic gain device to amplify low-level signals, and the two circuits together are known as a Compressor.

By limiting intense signals and amplifying the weaker, it is possible to compress any sound into a narrow dynamic range. This can be very useful if it is wanted to broadcast all the sound at the maximum possible volume as in the case of radio adverts and pop music records. The automatic recording device on many home recorders is a type of compressor, and is further discussed under the heading **Autorecord.**

Loudspeaker

For good reproduction from your tape-machine an extension loudspeaker is necessary, and in the case of stereophonic units there will be two speakers placed apart from each other. The speaker built into the recorder will usually be too small, and the machine's cabinet unsuitable for good reproduction. Naturally one is willing to sacrifice the best sound for the convenience of portability, but if you are using a cassette machine at home it will sound far superior when connected to a decently sized loudspeaker in a sealed wooden cabinet. The output of your replay amplifier is designed for a specific *impedance* of loudspeaker and it is important that only a speaker of the correct impedance is used. Whether it is 3, 15, 30 ohms or high impedance will be marked on the output socket of the machine, and on the loudspeaker.

A speaker unit should be capable of reproducing the full audio spectrum from the growl of the tuba to the ringing cymbal. A combination loudspeaker has two cones, a larger one to reproduce the bass notes and mid-frequencies, and a smaller inner cone to handle the treble notes. These combination units produce quite good results at normal volume, and enable the speaker cabinet to be kept to a convenient size for the average lounge.

For better quality, and particularly for accurate reproduction of very low frequencies such as the playing of a double-bass, there should be separate speakers in the cabinet to handle the low, medium and the high frequencies. Also in the cabinet is a filter device called a 'crossover network' to feed the correct frequency range to its own speaker.

Often a cabinet will contain a large speaker to cope with the bass frequencies, and a combination speaker for the medium and high range. At very high levels of volume, such as is used by pop groups, the speakers are housed in separate cabinets to prevent interference to each other. The tiny treble speaker known as a *tweeter* is incapable of handling very high levels of sound and a horn-type speaker must be used.

The placing of your speaker is most important. The treble tweeter speaker is highly directional so that it should be placed pointing at the listener's ear, and at ear height. The acoustics of the room in which the speaker is used will affect the reproduced sound, and this is fully explained under the heading **Playback.**

Magic Eyes Some tape-recorders use a Magic Eye for measuring the record volume level instead of a VU meter. Often the eye is very similar to the VU meter, with the glowing strip of the tube behaving exactly as the needle does in the meter. However, there are many variations on the Magic Eye system according to the manufacture of the machine, and the maker's handbook should be consulted to find the indication that is given by any particular eye when the recording level into the recorder is correct. Then the record volume control should be adjusted so that the Magic Eye lights up to this amount when the microphone is being spoken into.

Meters There are two types of recording-level meter in general use: the Volume Indicating Meter, or VU Meter for short, and the Peak Programme Meter, PPM. Most home machines are fitted with the VU meter which is calibrated in decibels, but the readings are only true for a steady frequency tone. There is usually an arbitrary calibration from 0 to 100 on the meter, and an area above 100 marked off in red. This red area indicates an over-recorded signal and the needle should normally waver just peaking to the 100 mark for a good recording.

The VU meter is difficult to read because the needle reacts so very rapidly to the signal. It will also give the same readings for a dramatic speech as for louder but less undulating pop music. With a little practice it is possible to understand the performance of the VU meter in conjunction with a monitoring loudspeaker, and produce perfectly balanced recordings. However, professional studios and tape machines utilise the PPM which is covered under the heading **Peak Programme Meter.**

Microphones The microphone is the ear of any sound system, it receives the sound-waves and converts them into electrical impulses. There are a number of different types of microphone, the more common being the crystal mic, the dynamic or moving-coil microphone, the condenser, and the ribbon.

The cheapest microphone is the crystal, and it is commonly used with the more inexpensive recorders. It gives a high output, but tends to accentuate the treble frequencies giving an imperfect recording. The crystal microphone can usually be replaced in your recorder by a dynamic microphone of the high impedance variety.

The moving-coil or dynamic microphone is an efficient unit suitable for all types of recording and is relatively cheap. The basic mic is sensitive to sound arriving from all directions, but many manufacturers now design their stick dynamic microphone with a special slotting system at the base of the stick which gives it a directional response. Its pick-up area is normally a heart-shaped pattern spreading out in front of the microphone making it receptive to sounds that the mic is pointed at, or which are produced close to it. Sounds that are produced behind the mic are not picked up, or are at a low level. If the dynamic microphone is placed very close to the mouth or to any sound source, the bass frequencies are accentuated. This can be used to record deep brown voices and is common practice on radio stations to give the announcers more authority.

The ribbon microphone is sensitive to sound arriving at two opposite sides giving it a response pattern that resembles a figure of eight. It has been used by the BBC as a general purpose studio microphone since the 'thirties, as two speakers can sit opposite each other and use the same microphone placed in between them. In the home the ribbon is now rarely used, being more fragile and expensive than the dynamic mic, and because it produces a rumbling if used in a draught or wind.

The condenser microphone is an expensive professional unit which can be switched to become sensitive to sound on either side, in front of the stick, or to pick up sound from all around.

Stereo recording requires two microphones of equal performance, so these should be of the same type and manufacture. It is possible to buy a stereo microphone which simply consists of two dynamic microphones, placed at right angles to each other, and housed in the same casing.

Many microphones are available with a switch on the stick so that the tape-machine can be switched on or off remotely. As long as you remember to turn the switch on when recording or hoping to record, then this is an excellent facility for when you are dictating or recording a discussion. Some microphones are fitted with a switch that simply switches the microphone off and does not stop the machine. I cannot see a purpose for this type of switch and believe it will cause trouble by being left in the off position when you are trying to record.

If you are about to buy a microphone I would advise a mid-price range dynamic microphone with a cardioid response, and fitted with a remote on/off switch for the recorder. It should be light-weight, but large enough so that you can get a firm grip on it. Should you wish to record sounds from all around the microphone, this dynamic can be strung up pointing downwards similar to an electric light. Although it will not have a true omni-directional pattern, in this position its heart-shaped response will take in sounds from all around beneath its position.

Microphones are delicate instruments and should not be dropped, shaken, blown into, or left so that dirt can collect inside the mesh.

Mixers

If at any time more than one input to your recorder is needed for producing drama tapes or animated lecture recordings, or even just a more exciting letter tape, then a sound mixer is required. This will enable you to play in music or sound effects from a record-player and fade it underneath your voice so that you are speaking over the effect or music. You can mix a previous recording with your own voice making a mock interview or perhaps answering your letter tape correspondent in person by mixing your voice with a snatch from his last tape. Disc jockey type tapes can be produced for replaying at parties by mixing your voice microphone with two record-players, each playing in a disc in turn. As discussed under the heading **Leader,** an insert tape can be compiled which can be mixed with music and live narration to produce a complicated sound production similar to taped radio programmes.

Some machines have a built-in mixer with the control faders set next to the tape deck. This system is very convenient as the microphones, radio or record-player can simply be connected to the input of the mixer and balanced by adjusting the faders, the output being connected directly to the recorder.

A simple mixer can be bought from most Hi-Fi shops to suit any home recorder, and an inexpensive unit with four inputs will be sufficient for most purposes. Two of the inputs should be low level for use with microphones, and two should be high level to handle record players or another tape machine to play-in insert tapes. The output of the mixer should be high level and of high impedance so that it can be directly wired to the 'auxiliary' or 'radio' input socket on the tape-recorder.

To obtain a good balance of sound it is useful to have a tape machine with a socket to take a pair of monitor headphones. This will enable you to listen to the various elements as you are adjusting the faders and produce a good 'mix' which is then recorded. If your machine does not have a monitor socket then by experimentation find the position on the faders where the various sound sources should be adjusted to produce a good blend of sound. These fader positions can then be marked. For example if a record is to be faded down and a voice recorded over the top of the music proceed as follows: Play the disc with its fader fully open and the record volume control on the machine set so that the VU meter has its needle moving up to the start of the red area on the dial. As the music comes towards the end of a phrase open the microphone fader and close the music fader by about one-third. Start speaking into the microphone keeping the distance between your mouth and the microphones constant. If the VU meter needle is swinging into the red area of the meter scale, close the mic fader slightly to bring it so that it just peaks to the start of the red area. Now mark the position of the microphone and record-player faders. If the tape is now played back the balance of sound can be assessed. It is probable that the music sounds too low under the voice in which case the record-player fader must be less closed when the next music fade is attempted. If the music is too high and obliterating the voice then at the next mix the record-player fader must be closed further than the mark now on the scale. When the correct position for the music fader is found this should be marked and your original experimental mark erased.

Multi-track Recording

Multi-track consists of gradually adding more and more to a recording until a complex production is achieved. In music studios 16- and 32-track machines using 2″ tape are quite common. The first track may be used to record the violins, the second for the choir, the third for the sound effect of seagulls, and so on. Each new track can be synchronised with the others as the heads are combined record and playback. This means that tracks one and two can be played into the ears of the musicians as they record the third track. Since each contribution is on a separate track, it can be erased and re-recorded until perfect without affecting previous recordings on the other tracks. This is the way pop records are built up, often with the singer harmonising with himself or simply singing twice to give the sound more character. When all the tracks are thought to be satisfactory, the filters and effects units, the echo and volume of each track is adjusted so that when played together they produce a pleasing sound. The tape is now copied to normal ¼″ tape. This final mixing process is known as 'tape reduction'.

At home it is possible to multi-track by laying one part of the recording on one machine, and then copying this on to another recorder while adding the next part of the production. This composite recording can then be copied back to the first machine while a third part is added. In this way it is possible to record a three-part harmony with one voice, or build up a three-voice comedy by yourself.

If you have a stereo machine it is quite easy to record one track only by switching just the first or right-hand channel to record. This can then be spooled back and played into the singer's headphones while she sings the harmony with the second or left-hand channel recording. Providing that the machine has combined replay/record heads, the two tracks will be synchronised and can be played back as a combined harmony.

Many domestic recorders are able to superimpose one signal on another. However, this system of over-recording cannot give good results as the second recording always has the effect of slightly erasing the first signal, particularly the treble frequencies.

Music

Music can be recorded by placing a microphone near every instrument that is playing and by balancing all the signals received from the battery of mics. This system is used in television where the studios are acoustically very dead, and the mixed signal is played back to each musician as he performs to compensate for the lack of reverberation. It is also necessary to use a large number of microphones where a large number of musicians are playing music with a wide dynamic range. During the quiet passages, such as a flute solo, the particular microphone near the soloist will be amplified while the others will be faded out at the mixer desk to avoid background noise.

With a pop group or brass band it is possible to walk in front of the stage and find a point at which all the instruments naturally blend well. If this is not the case then the band in question have arranged themselves badly. The louder instruments and the drums should be towards the rear. When the point of a *natural acoustic mix* has been found, a dynamic microphone can be placed in this position pointing towards the band, and a good recording will result.

To record a pianist, a fairly resonant room should be chosen. If the room has a carpet, roll it up as this will liven the acoustic. A dynamic microphone should be used fitted on a floor stand about three feet from the treble end of the piano, preferably pointing in the back of the instrument. If the pianist is to sing then ideally there should be two microphones used, one for the piano and the other placed near the singer's face. The output from the microphones should be fed to a mixer where the two signals can be balanced to provide a pleasing sound. Neither mic should be placed on top of the piano, and both should be as far away from the bass end of the keyboard as possible.

With only one microphone, the pianist must try to play softer or his voice will be drowned. In this case the microphone should be placed to the right of the singer's face away from the piano. Most other instruments can be easily moved to and from the microphone by the musician as he plays, so that the quiet passages are not lost, nor the crescendo over-recorded. By moving his guitar upwards and downwards a folk singer can easily balance his voice and the instrument to one microphone, and this arrangement adds colour to the recording since the 'off mic' guitar is more subdued than when the soundbox is placed up to the mic.

Certain instruments are highly directional in their output and the microphone must be placed so that it receives the full sound. The violin projects its sound at right angles to its soundbox, and the flute sends its notes spinning off the mouthpiece.

Noise

Any sound that is reproduced by the recorder but was not fed into the microphone has been originated in the machine itself and is unwanted noise. This is different from unwanted background sound which is picked up by the microphone. The latter can be excluded by recording in an insulated room, or at least one with the windows closed and the curtains drawn. In such conditions the only background sound will be from the recorder's motor, so the microphone should be placed well away from the machine and never on the same table as the recorder is operating.

The unwanted noise that is produced in the recording system itself can take the form of hum, hiss, whistling or ringing. Hum is usually caused by bad cable screening or close proximity to a mains cable, and its elimination is explained under the heading **Hum.** Similarly **Hiss** is tackled under its own heading. A whistling sound or a spurious tone appearing on the recording is caused by a nearby oscillator. This can be from a television set, a radio-paging system, a radio control unit for model aeroplanes, or a radio-telephone. The only course of action is to track down the offending piece of apparatus and turn it off. A ringing sound may appear on the recording if either the recorder or any ancillary equipment uses valves. By running the machine and tapping each valve in turn it will become apparent which is producing the ringing and it should be replaced.

Any recording will become noisy if the tape heads and guides are not regularly cleaned, or if old tape is used. But even with a properly operated recorder there will be some low-level hiss due to the system, but it is usually too low below the volume of the recorded sound to worry about. If a tape is copied to another tape, and maybe copied again, then the final tape will have three times the tape noise that the original had. It is in this situation that the noise level begins to threaten the quality of the recording and if the tape is to be copied to a third generation then a low noise tape should be used. There are also noise reduction systems available which are used in professional studios, the best-known of which is the Dolby. It is possible to buy tape-machines for domestic use with built-in Dolby amplifiers, and if these units are fitted to a cassette recorder using chromium dioxide tape, beautiful professional tapes can be recorded. See page 109.

Outdoor Recording

For location recording a portable machine is necessary, and should be used in conjunction with a dynamic moving-coil microphone. A ribbon mic should never be used as it is far too sensitive to air currents and will mask the recording with wind rumble. The recorder should be set down firmly on a wall or the ground to avoid wow and flutter. Even with cassette machines, the motor speed will vary if it is dangled from the shoulder or held vertically in the hand. It will only operate stably if set in its normal functioning position.

As normal, a pre-recording test should be carried out as fully described under the heading **Interviewing.** On playback of the tests there may well be some bass rumble or loud blasts on the tape. This is caused by air currents although it may seem as though there is scarcely a breeze. This is overcome by using a wind-shield made from sponge synthetic rubber. They are readily available on the market, but you can easily make your own by cutting a snug housing for your microphone in a block of foam rubber, using a narrow razor-knife. In emergencies a scarf or woollen glove can be wrapped around the microphone as a make-shift wind-shield.

When recording outdoors it is as important to consider the immediate acoustics and background sounds as when recording indoors. If your man-in-the-street is interviewed near a large building, the tape will sound like a man-in-the-church with added traffic noises. Similarly, the rural feel of a cow mooing will be lost if there is a transistor set playing in the background. The only real guide is to open your ears and really listen to how the place sounds.

See headings **Birdsong** and **Vox Pop.**

Pinch-wheel

The pinch-wheel is that hard rubber tyre that traps the tape hard against the capstan whenever the machine is switched to 'play' or 'record'. The rotating capstan runs against the pinch-wheel at constant speed and pulls the tape past the heads and on to the pick-up spool. If the recorder has been switched off at the mains but the tape deck left switched to 'record' or 'play', then the pinch-wheel will be left locked against the capstan. This results in a slight dent or flat on the rubber surface which will allow the tape to slightly slip during operation. For this reason it is always wise to check that the tape deck has been switched to the 'off' position when the machine is not in use, as only then will the pinch-wheel be disengaged and avoid wow and flutter. The ferric oxide coating on the tape gradually builds up a deposit on the pinch-wheel. This will vary the speed of the tape if it is not regularly wiped away with methylated spirits on a piece of cotton wool. Oil should never be applied to the pinch-wheel or to the capstan.

Playback There is little point in going to great lengths to secure a balanced and artistic recording if the tape is to be played back through a minute loudspeaker in a plastic box in the bathroom. To re-create the sound that was originally recorded, a properly enclosed loudspeaker must be used in a non-reverberant acoustic, such as a softly furnished and carpeted lounge. The speaker should be at ear level and pointing at the listener, not at a reflecting wall. The replay volume should be adjusted to reproduce the volume of the original sound that was recorded.

For stereo reproduction, the two loudspeakers need to be set well apart so that from where the listener is sitting the angle between the speakers is 60°. The listener and the two speakers will now be situated at the corners of an equilateral triangle. The two stereo channels should now be adjusted so that the correct three-dimensional sound is reproduced at the listening point. This does not necessarily mean that each speaker will be at the same volume.

Stereo headphones reproduce a pleasing sound, but it is not the same as using two speakers, and it is not stereophonic. This is because one channel alone is fed to either ear, whereas with two loudspeakers each ear receives something from both channels, and the reverberation of both around the room.

Many recorders and most replay amplifiers are fitted with tone controls so that the bass and treble can be accentuated. Depending on the quality of your speaker, the bass control will probably need to be turned up to reproduce the low notes of the recording faithfully. The treble control should need no adjustment from the normal with a good recording, and any treble accentuation will only increase the tape hiss.

Peak Programme Meter The PPM is fitted to more sophisticated recorders and studio apparatus to show the level of sound signal. Unlike the VU meter, the Peak Programme Meter is calibrated into equally spaced divisions, each representing a change in level of four decibels. The meter needle reacts immediately to any increase in signal level, but returns very slowly taking three seconds to fall the complete scale. In this manner the needle makes it easy to read the level of the peaks of the recorded sound.

The needle should be set to the centre calibration marked 4 when fed with that level of 1,000 Hertz tone that fully saturates the tape without distortion. This will be the 'zero level' of your recording system, and all other sound levels are expressed with relation to this fixed level, in decibels. See **Decibels.** Speech will now be fully recorded when it peaks between 5 and 6 on the PPM, and music should peak between 4½ and 5 to be fairly balanced.

Preamplifier

A preamplifier is an amplification unit that is placed between the output of a low-level signal source and the input of a normal amplifier. A microphone, for example, produces very low level signals and these need to be amplified before they can be fed to a speaker amplifier. It may be that in a particular unit designed for fitting between a loudspeaker and a microphone, the preamplifier and the power amplifier are combined, but the first circuit that the signals reach need to be low in noise level, efficient in operation and usually contain frequency adjustment devices. So, a *preamplifier* is used to boost the microphone input signals of a tape-recorder before they are fed to the recording amplifier. And the replay equalisation occurs during the replay head preamplification before the reproduced recorded signal is applied to the loudspeaker amplifier.

Most stereo enthusiasts will associate the term *preamplifier* with their box-of-tricks that adjusts the various sound sources before they are fed to the loudspeaker amplifiers. It normally has a switch to select tape-recorder, radio or record-player and tone controls. There will also be a control to adjust the volume of each channel, and perhaps another to mix one channel with the other. Here again the preamplifier is performing its role of boosting the chosen signal, and adjusting its frequency characteristics before feeding it on to the power amplifier. In this particular case there may be different impedance input circuits to accommodate different types of input equipment. There will usually be one or two high level input channels for signals that have already been boosted within their own circuitry. A high level signal will be attenuated within the preamplifier, then frequency adjusted and finally reamplified before being fed on to the power amplifier.

Preamplifiers sometimes take the form of *low level mixers*. In this design various signals can be mixed together before being frequency adjusted and amplified. In some recorders there is an input mixer which is this form of preamplifer. Signal sources from different microphones can be mixed with inputs from record-players or other tape machines and then passed on to the recording amplifier to be laid on the tape. Here again there may well be some high level inputs equipped with attenuators to handle signals from other amplification equipment.

Presence

In acoustic and sound recording the term 'presence' refers to that band of frequencies in which the human voice carries most of the language information, and to which the human ear is particularly sensitive. The peak of the ear's sensitivity is around 3,500 Hertz and it is this frequency in practice that is referred to as 'presence'.

When voices are singing with music that is equally loud, such as during an advertisement, they can be made clearly intelligible by accentuating the presence frequencies. A *presence unit* is commonly used in pop music and for disc jockeys to enable them to be heard above their records. Some microphones have been especially designed which are more sensitive at the presence frequencies to enable the user to penetrate background music.

Pressure Pads

Many home recorders are fitted with small felt pads that push the tape into the heads during operation. Although this ensures close contact between tape and heads, these machines are more difficult to edit on as it is impossible to mark the tape where it lies on the replay head. For this reason a good editing recorder is one which keeps the tape against the heads by tape tension alone, and without pressure pads.

With the pressure pad system it is imperative that the springs behind the pads are kept free from dirt, and the pads themselves are regularly cleaned. Failure in this respect will result in a loss of treble on recordings and possible dropouts.

Print-through

Print-through is the effect brought about by one turn of magnetised tape slightly magnetising the wind of tape beneath or above it on the spool. It produces a low-level copy of itself on the adjacent winding which can seriously mar the original recording. Heavy print-through is more likely to occur with thinner tapes such as double-play or triple-play.

With modern tape it is possible to record at quite high levels without distortion; however, a highly recorded tape is more likely to suffer from print-through. The effect becomes more significant the longer a recording is kept, and it is advisable to play through archive tapes so that they are rewound in a different placing on the spool.

Quality

The quality of a recording is an assessment of how accurately the system can pick up, retain, and reproduce sound. A high quality recording is one that when replayed contains all the frequencies that were contained in the original sound to the correct degree, and does not include any extra noise or coloration. Providing that there are no inferior components in the electronics, the transistorised tape-recorder will faithfully record and reproduce sound at high quality. Higher quality will be obtained at the higher tape speeds, as this increases the frequency response of the system. Radio studios commonly use 7½″ per second for speech and music. Quality will suffer if there is dirt on the tape heads and guides, or if a portable machine is being run on tired batteries. Do not use old tape as it will cause trouble to the machine and produce bad recordings.

Most modern recorders will give excellent results if kept clean and fitted with a good microphone and loudspeaker. So often a sound system produces bad results because of a poor or damaged microphone.

Quadrophonics

Quadrophonics is a four-track, four-channel stereophonic recording system. It is capable of recording separately the outputs from four microphones placed at right angles to each other. The four tracks can then be reproduced through four loudspeakers set in a square around the listener, and the complete audio atmosphere of the original recording studio is re-created.

Although there are four-track, four-channel machines on the market using normal ¼″ tape, there are at present no commercial recordings in quadrophonics. A four-channel record-player can be devised using a coding signal for converting the signals into a two-track recording which can be easily cut on to a disc. When the record is replayed the output is decoded back into four channels and fed to the speakers. Presumably because of the cost involved and the lack of interest from the public, the manufacturers are slow in developing a domestic quadrophonic system. However, quadrophonics is being used for cinema sound-track, and by audio is able to convince the viewer of what is on the screen far better than the three-dimensional visual experiments of the 'fifties.

If you would like to try four-channel recording at home, borrow a friend's stereo-recorder and link it with your own. The room that is to be used as the studio should be acoustically prepared as much as possible, since you will be recording from all directions. Set up the two stereo microphones back-to-back so that they cover the whole room. If you are using separate mics, set these in a cross. The microphone stand should be placed in the centre of the room, and the performance or drama conducted naturally, only filling the whole room. It is a good idea to bring plenty of movement into whatever you are recording, so that the qualities of quadrophonics can be enjoyed. But remember, your actor will only be heard to be moving if he is speaking at the time!

The two tape-machines should be placed as far from the microphones as is possible. Each should be adjusted for recording level as described under the heading **Stereo.** Both machines should be switched to record, and before the performance starts count loudly from one to five. On playback the two tapes will need to be synchronised, and by pressing the pause button on one of the machines after the count of three, and then playing the second machine, the first tape can be set in motion as the count of three comes out.

Radio to Tape Most transistor sets have a socket marked 'tape', and this output provides a properly balanced feed for recording. This socket should be connected to the recorder input marked 'radio' or 'auxiliary' using the appropriate plugs and lead. You may well have a connector that was supplied with the tape-recorder, but if not, try a large Hi-Fi shop. With older radio sets there will not be a tape socket, and the feed must be taken from the socket or terminals marked 'extension loudspeaker'. This output from the radio will be at higher level, but it can be controlled by the radio's volume control which should be turned down low.

To obtain a good recording of the programme, the radio station must be tuned-in properly. If it is a transistor set then it should be rotated until the position of maximum reception is found. An aerial should be used if the station is on VHF or the shortwaves, and many sets have an extending whip antenna which gives good results. Connect the radio to the recorder, and with both switched on adjust the station tuning control until the received signal gives the highest reading on the VU meter. Now the record volume control can be adjusted so that the needle of the meter just swings up to the red part of the dial. The programme will now be recorded clearly, and if the station is operating on VHF, the quality can be excellent.

Trouble may occur when recording from a stereo radio as the multiplex transmission system signals of very high frequency may interfere with the bias signal of the recorder. The solution is to fit a cut-off filter at about 20 kHz between the radio and the recorder, and your specialist stereo shop will be able to supply this and the inter-connecting leads.

Sometimes it is not possible to connect the radio receiver and the tape-recorder because of bad design or lack of a connector. In this case the only solution is to place the microphone in front of the radio loudspeaker. It should be pointing into the centre of the speaker, and just four inches away from it. If it is placed any farther from the speaker then you will be recording the atmosphere and background sounds of the room. The station should be tuned-in clearly, and the volume control on the radio turned up to a reasonable level. The record volume control should now be adjusted so that the needle of the VU meter swings just up to the red area on the dial. This will be a lower setting than normal because the microphone is so close to a fairly loud source of sound. During the recording of the programme the room should be kept as quiet as possible since the microphone will pick up any loud background sound besides the radio.

Having encouraged you to record from the radio, can I now direct you to the heading **Recording Restrictions.**

Record

Here is a step-by-step procedure to ensure a good recording. Choose a softly furnished room away from the street, close the door, the windows, and draw the curtains. Arrange an armchair in the centre of the room facing towards a corner. Put the tape-recorder on the floor to one side of the chair so that the microphone can be held when sitting down, but with the mic cable fully extended.

Place a full spool of tape on the left-hand mounting of the tape-machine. Lace the end of the tape through the tape guides, across the heads, between the pinch-wheel and the capstan, and on to an empty spool on the right-hand reel mounting. Make sure that the heads are clean and that the tape has its rougher oxide surface facing in towards the heads.

Plug the microphone into the machine and if there is a switch on the mic stick switch it on. Plug the recorder into the mains, or if it is a battery model see that they are new. There is usually a test switch and meter to read the strength of the batteries. Switch the machine on.

Your machine may have an input switch marked Mic/Radio/Auxiliary which should be switched to the Mic position.

Now switch the machine to record. Holding the microphone 9″ from the face, recite a nursery rhyme and adjust the record volume control so that the needle of the VU meter moves up to the red area on the dial, but not into it. Now recite another nursery rhyme.

Stop the machine, spool the tape back to its beginning, and play the recording. Listen carefully to the second nursery rhyme which was recorded with the volume properly adjusted. If the sound is distorted then the record volume control was set too high. Maybe you were holding the microphone too close to your face and blasting it. You will also notice whether the room you have chosen is too echoey, or if there is too much background noise.

Having resolved any difficulties, wind the tape back to the beginning, switch it to record, and settle back in the armchair with the microphone held at 9″ away, pointing towards your mouth. You can now start speaking, but if you are using notes or a script be careful not to hold this directly in front of your face, or between the microphone and your mouth. If the paper is laid flat on your lap it will not reflect the sound, and can be easily slid on the floor revealing the sheet under it without the noise caused by turning over a page.

Perhaps when you come to play back the recording, there is nothing on the tape. This may be because the microphone switch has been knocked off, because the tape has been threaded incorrectly or is inside-out, or because you switched the machine to 'Playback' instead of 'Record'.

This procedure for a good recording is identical for cassette machines, only with these the tape is helped into place and threaded across the heads by its cassette housing. With stereo machines there will be two microphones or a combination mic, two record volume controls, two meters and possibly two input switches. The setting of the two volume controls for a three-dimensional effect is described under the heading **Stereo.**

Recording Restrictions

Many sounds that you may wish to record will be the property of the performer, the writer, the composer or perhaps a company that has bought the rights. This will not matter if you are only recording for fun, but technically you are infringing copyright if you make an unauthorised recording of any published musical work or literature. If you are seen recording a prominent artiste at a concert your tape-machine could well be confiscated since performers lose a large amount of income to bootleg record manufacturers. Very often these pirate records are made from tapes recorded at public concerts.

If you intend using copyright material in any way publicly, even if it is only a snatch of music on a lecture tape, then permission should be sought from the appropriate body and the small fee paid. Here are the addresses of the copyright protection bodies in Britain:

Phonographic Performance Ltd,
62 Oxford Street,
London W1.

The Performing Right Society,
29 Berners Street,
London W1.

British Copyright Protection Association,
50 Newman Street,
London W1.

British Actors' Equity Association,
8 Harley Street,
London W1.

Mechanical Copyright Protection Society Ltd,
380 Streatham High Road,
London SW16.

The Musicians' Union,
29 Catherine Place,
London SW1.

The British Broadcasting Corporation,
Broadcasting House,
London W1.

The Independent Broadcasting Authority,
70 Brompton Road,
London SW3.

National Union of Journalists,
314 Grays Inn Road,
London WC1.

The Society of Authors, Playwrights and Composers Inc.,
84 Drayton Gardens,
London SW10.

The League of Dramatists,
84 Drayton Gardens,
London SW10.

Copyright infringement is particularly serious if your tapes are to be published, broadcast or played to a large audience. Should you be making recordings for a college or a club then any live music or copies from discs should be paid for. The Mechanical Copyright Protection Society will give you advice on the telephone if you are unsure of which copyright body to approach, and will indicate the cost of using a recording for particular purposes.

Here is a summary of the 1956 Copyright Act supplied by the Society:

1. It is an infringement of copyright to make a recording of copyright music or any other copyright material for any purpose whatsoever without the consent of the owner or his agent.

2. It is only permissible to record the programmes of the BBC and IBA for private use but this does not extend to any copyright material included, and such recordings

would infringe the copyright unless prior permission has been obtained from the owner or his agent.

3. If permission is given to record copyright music at home, it is usual to restrict this to private and domestic use only, and permission for public performance, playback or broadcasting is withheld except in very special circumstances.

4. It is an offence under the provisions of the Dramatic and Musical Performers Protection Act 1925, as amended by the Copyright Act 1956, to record the performance of an artiste without his written consent.

5. It is not permissible to record an amateur dramatic performance or any other public performance involving copyright material, and a clause forbidding this is usually inserted into the licence given by the copyright owner to the amateur dramatic company whereby they are authorised to perform the work in public.

Scriptwriting

The art of the scriptwriter is to lay on paper a simulation of how people really speak. This means the sentences must be kept short and familiar phrases used. People rarely use long words in everyday speech, but utilise colloquial terms and expressions to communicate attractively. It is this attractive quality of natural speech that the scriptwriter must capture, so that the narration keeps the attention of the listener as well as conveying information.

When the script is to be read by yourself it is possible to personalise it, bringing in your own pet catch phrases. The best way to write your own script is to put down, word for word, the way you would tell the story or information to a friend. Remember that we use 'don't', 'there's' and 'can't' in speech rather than the full form.

Reading to the microphone is very different from speech-making, and care should be taken not to confuse the two. A speech is directed at a large audience listening at the same time, and who are visually focused on the speaker. A tape is normally heard by a few people at a time, and the words coming from the loudspeaker should sound warm and personal to compensate for the loss of visual communication. By using tape, teachers are often able to get through to students that do not react to the traditional lecture. This may be because the tape is a novelty, but more likely that the teacher has been able to project himself and his ideas better than when he has been lecturing in front of a large number of students.

When the script has been written, carefully read through changing any difficult words for those that flow off your tongue. Make sure that the sentences are easily understood when read aloud. Long rambling sentences should be cut up into short statements. The knack of reading the script is to sound as though you are simply talking and not reading your words. Practise is the only way to learn, and by recording your attempts and playing the tape back you will see how well you are faring. Often by trying to put meaning into the delivery, a narrator will emphasise the wrong words, or even worse, all the words. Underline the words that need stress in your script, and try to get your delivery flowing naturally. Take plenty of breaths, keep the pace slow and relaxed, and try to speak the way that you talk to your family and friends. Even professional personality broadcasters can become boring if their script is unbroken for too long. It is better to speak for two minutes and then illustrate some point with music or an interview than to rattle along for five minutes uninterrupted.

Speed

Tape-recorders commonly have three speeds to which they can be switched, 1⅞" per second; 3¾" per second; and 7½" per second. This corresponds to 4.75, 9.5 and 19 centimetres per second.

It would appear that the lowest speed was the best to use for any purpose as it economises on tape. For example a 7" reel of normal thickness tape recorded on half-track at 1⅞" per second will last for four and a quarter hours, while at 7½" per second it will last for only one hour. The reason for using the less economic speeds in terms of tape, is that the frequency response of the system is better and a higher quality recording results. With modern machines and tape it is possible to compromise, and 3¾" per second will give satisfactory results for speech and music. If it is a recording that will be edited later then 7½" per second should be chosen, as this assists in easily tape marking the edits.

The low speed of 1⅞" per second is excellent for storing a lot of information on a small amount of tape. This speed can be used for dictating and letter tapes, and whenever the quality is not a prime consideration. That is not to say *all* low speed recordings are low quality because cassette machines and crossfield head recorders achieve very good results at low speed. In general, however, the highest possible speed on your machine will give the best results, and will be least affected by any creases or dropouts in the tape.

Spool sizes and tape duration at 3¾" per second for single track

Spool Size (*the diameter*) *in inches*)	Long Play Tape (*feet*)	Duration (*minutes*)
3	210	11
4	450	22
5	900	45
5¾	1200	60
7	1800	90
10	3600	180

Spool Size	Double Play Tape	Duration
3	300	15
4	600	30
5	1200	60
5¾	1800	90
7	2400	120
10	4800	240

Spool Size	Triple Play Tape	Duration
3	450	22
4	900	45
5	1800	90
5¾	2400	120
7	3600	180
10	7200	360

Recording time in minutes per track

Tape length (*in feet*)	Tape Speed 7½ ips 19 cms/sec	3¾ ips 8.5 cms/sec	1⅞ ips 4.25 cms/sec	¹⁵⁄₁₆ ips 2.125 cms/sec
210	5.5	11	22	45
300	7.5	15	30	60
450	11	22	45	90
600	15	30	60	120
900	22	45	90	180
1200	30	60	120	240
1800	45	90	180	360
2400	60	120	240	480
3600	90	180	360	720

Spooling

To wind the tape tightly and neatly on to another spool, arrange a pencil by the take-up reel so that it slightly bends the tape at an angle to the vertical. Then when the machine is switched to 'spool' or 'rewind' the tape will enter the reel at that angle which helps the air to escape from beneath the winding, and gives a good tight reel of tape. This is particularly useful when sending letter tapes, and the maximum amount of tape can be spooled in this manner on to a 3″ reel.

When spooling through the tape during editing, be careful not to suddenly switch the tape deck into the stop position. This will immediately brake the machine and the tape will stretch or snap. It is better to slow the spooling action by switching the tape deck to spool in the opposite direction. This will cause the tape to slow down smoothly, and at the point where it is almost stationary the machine can be switched off.

Stereophonic Recording

There is no mystery to stereo recording and reproduction even though many enthusiasts turn the subject into a secret rite. Stereo tape-recording is simply making two *synchronous* recordings on the same tape from two microphones placed at right angles to each other. The two feeds of sound are known as *channels* and they are differentiated by calling one the right or first channel, and the other the left or second channel. So that the two channels are recorded simultaneously and separately there is a separate record volume control, amplifier and recording head for each. The two recording heads are mounted immediately one above the other so that they lay two separate narrow tracks side-by-side on the tape. As the two microphones are set at right angles to each other they will be picking up sound from different parts of the room. Some of the sound will be received by both microphones of course, but the two channels represent two separate ears full of sound. On replay, one channel is fed to its own loudspeaker set some distance from the second speaker serving the other channel. If the listener now stands so that he forms an equilateral triangle with the two speakers, he will hear a three-dimensional sound that cannot be produced by a mono recording.

A stereo microphone is two microphone units conveniently housed in one casing at right angles to each other. If two separate microphones are used they should be of the same type and clamped at right angles to each other on a floor-stand. With the microphones set in place pointing over the studio, the two record volume controls should be adjusted. The sound source, perhaps yourself reciting a nursery rhyme, should be positioned on the line between the microphones, and the two record volume controls adjusted so that the needles in the two meters swing up towards the red areas on their dials. When the sound sources are spread out in the room the meters will have different readings as the levels received on the two channels will be different. This is how things should be, and the record volume controls should not be re-adjusted to give equal readings. See headings **Recording** and **Replay.**

Superimpose Superimposed recording is covered under the heading **Multi-track Recording.** If your machine does not have a superimpose switch and you wish to over-record, not erasing the original recording, place a piece of cardboard so that it keeps the tape away from the erase head. This is the first head that the tape passes, and it is possible to arrange the cardboard so that it keeps clear of erasure but is fed past the record head. The original recording will be lowered in level beneath the superimposed signal, and will suffer in quality. If the two recordings need to be the same volume then the first must be made at maximum volume, and the second made 6 dB lower, that is half the volume. When both have been recorded the first signal will have been partially erased so that the two takes are equal in volume.

Tape

A different type of tape should be chosen to suit the task in hand. If a continuous music supply is needed then triple-play tape will provide a long playing time before the spool needs to be changed. For production and editing work standard play tape is required, as this will not stretch when pulled about during splicing.

The new low-noise tapes including chromium dioxide are excellent, but expensive. These tapes require the re-adjustment of the recorder's bias level which is not always possible on home machines.

Twin-track

See heading **Half-track.**

Uses

The domestic recorder can be utilised whenever information needs to be collected and reproduced at a later time. Letter tapes, drama tapes, film tapes, music, and radio-to-tape copying are covered under their own headings. Other uses include dictation; recording of telephone messages and conversations; step-by-step recipe recordings; language tapes; piped music; narration for slide shows; and as a teaching aid.

For dictating it is necessary for the machine to be fitted with a long microphone lead and a microphone that has a remote on/off switch. The typist will find it easier to replay the tape if there is a remote on/off switch of the foot-operated variety. She will also need a pair of headphones so that the sound is kept to herself. It is possible to dictate on to a tape cassettee and mail this to your secretary providing that she is given a written guide to any technical words or place names, and of course has a cassette machine herself. It is better that the dictation is given at a slow pace so that the secretary can type it without continually stopping the tape and winding it back.

The recording of telephone conversations smacks of bugging, but it can be very useful when receiving long orders, or perhaps conducting an interview over the phone. Older telephones can be simply fitted with a suction pad telephone pick-up coil, which is wired to the microphone input on the recorder. This coil receives an induced signal from inside the telephone and avoids any special taping to be made. The new-style telephones need a special recording socket to be fitted by the Post Office. Once this socket is provided it is possible to record from the telephone, and at a very reasonable cost hire a message machine from the Post Office. This is a simple recorder on which you place your announcement such as the time at which you will be back home to accept calls. When the telephone rings in your absence, this machine will answer the call with your recorded message. Other more complex telephone answering recorders are available on hire from private companies that will give your announcement and then record a message from your caller. Needless to say these machines are very expensive to hire.

Recipe recordings are an interesting way to cook better with confidence. Your library can be built by asking your more gastronomic friends to record on tape a step-by-step description of how they prepare their favourite dish. These recipe tapes, or cassettes, can then be played back in the kitchen as you have the ingredients to hand. If your friend has included every detail in the preparation, then the dish should be easily duplicated. Recipes are regularly read out on the radio, and these can be added to your library at your discretion.

Self-education tapes are an excellent way to learn a new language, since language consists of sound which is difficult to convey in writing. A tape and an illustrated book will often give the student a better grasp of the colloquial form than a teacher who has English as his mother tongue. Courses in almost every language are available on tape or cassette, and are advertised in the Sunday press.

A lot of pleasure can be derived by producing special music tapes to produce different atmospheres. A number of party tapes can be recorded with hard rock music to break the ice, reggae and soul for dancing, and more romantic records to end the evening. The technique of producing these tapes is covered under the heading **Disc to Tape Copying.**

A slide projection show can become far more entertaining and informative if a narration tape is produced to play with the slides. For the tape to make sense first the slides must be arranged in a definite order that conveys visually a certain theme. For example, your collection of shots of churches can be shuffled into order comparing and contrasting the architecture. Similarly your colour slides taken on holiday can be arranged to tell the story of the journey into foreign parts. See **Film Tapes.**

A script can now be written to complement the pictures, developing the theme and explaining whatever is shown on the screen. There are some tips given on successful writing and reading for the tape under the heading **Scriptwriting.** The narration can be animated by the inclusion of sound effects and music on the recording. The technical aspects of blending these sounds in with the voice is covered under the heading **Mixer.**

The use of music and effects will give life to the slides, break the monotony of a long narration, and also provide the cue for the next slide to be switched on to the screen. It can be arranged that when the carnival music starts to play on the tape, the next slide of a fun fair is brought on to the screen. After about fifteen seconds, the music fades under the narration describing how your family saw the complete fair erected in two hours. Similarly with effects: the sound of seagulls and sea lapping would cue the shot of your holiday beach at Corfu, and the voice would soon appear on the tape describing the geographic location of this particular stretch of coastline.

It is possible to buy special slide projectors that switch

to the next picture on receiving a pulse tone from the tape-machine. But this requires a recorder with a separate track available for sound synchronisation. There are slide-sound couplers available on the market which can be fitted on any half-track or four-track machine. They consist of a head unit and a pulse generator, and a replay amplifier. The coupler is fitted to the side of the recorder and the narration tape played through. At any point where a new slide should be shown, the pulse generator button is depressed. This records a blip of tone on the very bottom of the tape away from the first soundtrack of a half-track or four-track system. When the tape is played through again and the coupler connected to the remote control of the slide projector, the tone blips will be read from the tape activating the slide change. Because the coupler is using the bottom part of the tape, the narration must be recorded on one track only of the tape, that is to say when sound is put on the tape it can be run the whole of its length but not turned over and recorded back on the other side.

As with adding sound to moving pictures, the loudspeaker should be mounted behind the screen, or in the case of stereo either side of the screen. A well-produced animated narration in stereo will turn a slide show into real entertainment.

As a teaching aid the tape-recorder has been used for many years but usually as a tool for recording and playing back BBC educational programmes. It is possible to produce an animated tape complete with music and sound effects to replace the ordinary lecture. Interviews can be included, recorded between teachers, and this will keep the students awake and interested. The tape will come to life as a serious competitor to the television lectures if pictures are added by connecting the recorder to a slide projector. Teaching only succeeds if the teacher is able to convey his ideas and information to the students, and this is only possible when they are interested and receptive. The time spent producing effective teaching tapes and slides will be well rewarded, and there is no reason why the students should not be brought in on the recording and photography. Unfortunately many of the tapes and slides available ready-made for teachers lack the sparkle of showmanship and animation, and therefore fail in communicating their information.

Videotape Operating on the same principle as the sound tape-recorder storing signals from the microphone, the videotape recorder records electronic signals from a television camera. These can then be replayed into a television monitor where they build up a moving picture on the screen. A portion of the width of the videotape is normally used for audio recording, so that the picture is always synchronised with its sound.

Professional videotape recorders use 1″ or 2″ tape, but there are now available helical scan machines using ½″ tape and are more suitable for the amateur. Such a portable machine is capable of recording the picture and sound from a hand-held camera and mic, and instantly playing it back through the monitor in the viewfinder of the camera. If the shot is unsatisfactory then the tape can be erased and the piece re-recorded. Because there is no picture processing to delay the playback, amateur television is bound to supersede home movies within the next decade. The only present objection of videotape recording is the initial cost of the machine and of the tape. However, already there are Japanese recorders on the market that produce good results, can be connected to an ordinary television receiver for playback, are available in colour, and are within the range of many enthusiasts' pockets.

Vox Pop Vox pop is the term given to a recording containing many different opinions from the man-in-the-street. Each section of the production is kept as brief as possible so that the listener can compare and contrast the statements made, and the type of people that hold a certain opinion. Not all topics lend themselves to the Vox Pop technique since the large majority of the members of the public hold no views about an amazingly large range of affairs. However, if your subject is easily digested by the everyday person then Vox Pop can be very entertaining and informative.

Let us consider producing a tape on the theme 'Gentlemen prefer blondes'. Firstly, you must have a portable recorder. This can be a cassette machine if you are prepared to copy the cassette to ordinary tape later, for editing. You should make sure that your machine is loaded with a full reel of tape, and that the batteries are new.

Now you must position yourself in a public place where all types of person are constantly passing. It should be away from echoey passages and heavy traffic. There should be something convenient for you to rest the tape recorder on. A good location might be near a cinema queue or the exit of a railway station.

Once set, a test recording should be made to see that the machine is operating correctly. The procedure for this is explained under the headings **Interviewing** and **Outdoor Recording.** After the test, the machine can be set to record and the pause button depressed until you have lined-up an interviewee. If your machine does not have a pause button then it must be turned to record and off again before and after each recording.

Pick your interviewees so that you cover many different types of people: a businessman; a student; a career woman; a teenage girl, etc. Each type of person looks different, sounds different, and hopefully has something different to say. Politely stop your victim and ask if he or she would give their views on whether gentlemen prefer blondes. Naturally you will be stopping blondes, brunettes, and gentlemen that prefer neither, so you may well receive humorous answers.

When you start to record, behave as though you were approaching your interviewee for the first time, for example:'Excuse me sir, but I see your lovely wife here has golden hair, do you prefer blondes to brunettes?" For each interview the opening gambit should be varied, and in some way try to describe the interviewee. After taping about a dozen good interviews, return home for the editing. Eight different comments is enough to digest at any one time, and these must be chosen from the dozen or so recorded. They should be picked to cover a wide cross-section of opinion, and so that the more interesting accents and voices are used. Next they should be arranged in order so that each contrasts with its neighbours in voice quality and comment. Many of your own opening questions can be cut out so that the different voices speak one after the other. Each interview should be severely cut down to under half a minute. Only when the topic of gentlemen preferring blondes appears to have been lost introduce one of your opening questions. Providing that your speakers were characters and your editing efficient, then your Vox Pop tape, now lasting some four minutes, should encapsulate precisely the feeling about whether or not gentlemen prefer blondes!

Volume

The volume of any particular sound needs to be assessed before it can be efficiently recorded. If the sound is soft on the ear, then the microphone should be placed close to the source. At a distance it would be necessary to set the record volume control at a high setting which would also amplify any noise or reverberation in the room. With a loud sound the microphone should be placed away from the source so that the signal does not overload the pre-amplifier in the recorder. However, it should not be placed so far away that it picks up more reverberation than direct sound. The record volume control on the machine will need to be set at a low level with a loud noise, and the precise point will be indicated by the VU meter.

Dependent on the volume of a sound, and the position of the listener's ear relative to the sound source, is the loudness of the sound. Loudness is the degree of sensation experienced by the listener and it is measured in *phons*. But the human ear is more sensitive to some frequencies than others, so the loudness of any particular sound, when compared with another, will depend on the frequencies that make up the two sounds, and the power of the sound-waves by the time they have travelled to the listener.

At the threshold of hearing when a sound is barely audible the loudness is a small fraction of a phon. At the threshold of pain, when the sound is so loud it begins to hurt, the loudness is around 140 phons. But it all depends on the frequency of the sound. At 1,000 Hertz there is a separation of 140 decibels between the threshold of hearing and the threshold of pain. The loudness of any frequency sound is directly compared with this standard frequency tone, and when the ear hears both at the same loudness then the sound intensity of the tone is measured in decibels. The equivalent number expressed in phons is the loudness of the sound that has been compared against tone. See heading **Decibels.**

VU Meter See heading **Meters.**

Wow Wow is the slight changes of frequency of the sound on a tape produced by slow variations in the tape speed. It is particularly noticeable on sustained music notes when the pitch will be heard to vary. It can be caused by dirt on the tape deck, incorrect tape tension, or a worn pinch-wheel. If a tape has been spliced badly with the jointing tape off-centre on the width of the tape, the adhesive will prevent the turns of tape leaving the spool cleanly, and produce wow on reproduction.

Your Ear and The Microphone Our hearing system has many capabilities. It can pick up tiny sounds and immediately adjust to receive the loudest noise. In conjunction with the brain, the ear is able to be selective. It is possible in the middle of a crowded party to hear your own name being whispered on the other side of the room. We are also able to tell from which direction a sound is coming, and if we have experienced this particular sound before, we can estimate the distance of the sound source.

It is important to remember these properties of the ear when recording sound. The tape-machine cannot adjust to different level inputs unless it is fitted with an automatic recording device. This will enable the sounds to be technically recorded but it has the effect of ruining any audio perspective, as all the sounds are brought down or amplified to exactly the same level. A compromise must be made for good results: the different volume sounds must be brought within the dynamic range of the tape system by careful adjustment of the record volume control. Now when the tape is played back through a loudspeaker, the ear is given some indication that the original sounds were of differing level.

The microphone cannot pick out a particular sound for attention in the same way as the ear. When your name was whispered across the room at the party, if your ear had become a microphone connected to a tape-recorder, then on replaying the tape nothing would be heard. The sounds adjacent to the microphone would be at such a high level compared with the whisper that they would prevent it from being recorded. In more practical terms, this means that a particular room may sound acoustically pleasant simply because you are ignoring the echo and traffic noise from outside. Your ear is selecting whatever is being said in the room and discarding the background noise. However, when a tape recording is made in the same room these extraneous sounds will become apparent on replaying the tape.

The microphone can be selective in that it may have a directional response. But since the ear normally receives sounds from all directions, this property must be used to focus the recording on those sounds that are important without destroying the audio atmosphere. So in the case of our mythical party, it would be possible to record the far-off whisper by using a directional mic pointing at the gossiper.

To ensure a good recording, the microphone is often placed in positions unusual for the ear, and this will be noticeable on playback. A narrator may speak into a mic placed close to his mouth to avoid recording high background noise. On reproduction you will be able to hear his voice as though he were speaking directly into your ear, and containing heavy breaths, the clicking of false teeth or the smacking of lips. By putting the microphone close to certain sound sources it is possible to fool the ear into believing it is hearing something quite different, but from a distance. This is used in creating sound effects. For example, by screwing up Cellophane paper close to the mic, on reproduction the ear will believe this is the noise of a forest fire! See **Effects.**

Zero Level The volume of the electrical signal fed to the recorder by the microphone is expressed in decibels with reference to a standard level. This standard reference level is known as zero level. The zero level for a particular recording system would be suitably that level that provides a high level of recording on the tape without causing excessive print-through. But this is no set figure since types of recorder vary, and different types of tape may demand the use of a different zero level.

Section Three

The cassette recorder: its uses; how it operates; maintenance; allied equipment

The tape cassette: how it operates; various types available, their advantages and disadvantages

Chromium dioxide tapes

The Dolby system

The radio studio: its equipment and uses

The Cassette Recorder

The *Compact Cassette* tape-recorder has brought sound recording into almost every home. It has removed the need for technical knowledge to operate a recording machine and has made serious inroads into the position of the record-player. Cassette machines are cheap, light and portable, easy to use, amazingly good quality, and fulfil the public's demand for an instant sound reproduction 'black-box'.

The cassette machine has developed in all aspects. There are now very high-quality stereo machines, combined radio and tape units, tiny reporter recorders, mains or battery models, and designs particularly convenient for use in cars. All machines have the advantage of being small and easily powered by batteries. Unlike reel-to-reel tape recorders, the cassette unit needs no 'lacing-up'. The convenient tape package is simply popped into the appropriate slot on the machine and the 'operate' button pressed. The car-type models require even less, since simply pushing the cassette into the front slot automatically starts the drive motor. Again, unlike the traditional tape deck, the cassette does not have to be respooled to be taken off the machine. And, like a disc record, the cassette can be labelled and easily stored and recognised. For correspondence, the cassette is lighter than a tape and therefore cheaper to mail, and throughout the world there is a standard tape speed of 1⅞″ per second ensuring ease of playback wherever the cassette is sent.

Although there are very many models on the market, they all have a similar tape-drive mechanism and the varying electronic facilities available on different machines are to suit different needs. The simplest unit has a space to fit the cassette package, and capstan and spool motor shafts to take the tape across a replay head when the appropriate switch is thrown. A transistor amplifier boosts the signal to feed a small loudspeaker in the machine, and that is the complete unit. This type of cassette player is available at a very low price and is only capable of playing back pre-recorded cassettes, such as those on sale by the commercial gramophone companies containing music traditionally available on long-playing records.

The basic cassette recorder will contain a deck, switches for playback, record, fast spool forward, and fast spool backwards, a volume control, and a microphone fitted with a switch to stop and start the recording. There will also be a loudspeaker in the plastic case. In this type of machine, the recording level is adjusted automatically, which results in recordings that can sound 'edgey'. There are three methods of fitting the cassette easily on to the drive deck, depending on the model. The more basic recorders just have the drive shaft and capstan, and the two locating spindles protruding from the case of the machine where there is space to fit the cassette. The cassette is then fitted over the protrusions, with the tape gap of the cassette facing towards the front of the machine where the heads are situated. A neater arrangement, commonly employed, is a lid fitting over the drive deck. Along the sides of the lid are two brackets so the tape cassette can be slid directly underneath the lid, and coupled to it. The tape gap is arranged so that it faces outwards. When the lid is closed down on the deck, the cassette is automatically fed on to the locating pins, drive motor spindles and capstan. A third method is employed on those machines that have a letter-box type opening on one side of the unit, and are particularly convenient in cars. The cassette is pushed firmly into the slot with the tape gap facing into the machine. Inside a mechanism brings the drive deck laterally into contact with the cassette, and also automatically starts the machine on playback. For recording, this type of unit has a red button that must be depressed when pushing in the cassette.

These simple machines are practical and for replaying

pre-recorded tapes or recording correspondence are ideal. However, if good quality recordings are required, then the recorder should possess a record volume control and a volume indication meter. Often the replay volume control will double as the record volume control, and this is quite acceptable. The procedure for adjusting and recording with such a machine is fully explained in the A to Z part of this book under the heading **Record.**

It is possible to purchase high quality machines that contain Dolby units, which record and replay stereo. The same cassettes can be used for stereo as for mono, and most pre-recorded music cassettes are in stereo. A good recorder may well have a switch to change from standard cassettes to the low noise Chromium Dioxide cassettes. This facility is invaluable if you are considering recording your own music cassettes, since the new Chromium Dioxide tape gives much better music reproduction.

Before buying the more expensive cassette recorders, it is worthwhile considering very carefully the reasons why you require a recorder. If it is simply to reproduce music cassettes in stereo, then possibly a less expensive machine without Dolby units will suffice. If it is to record or replay letter cassettes, then probably a less than £20 model would perform the task adequately. If the idea is to record music albums to cassette, then the money would be better spent in buying music cassettes direct from the gramophone companies instead of LPs. Should you be buying a machine to record documentary, programme or lecture types of material, then the reel-to-reel recorder would be a far better choice. It is practically impossible to edit cassette tape, although there is a kit on the market. It is very difficult to splice the tape together and, even when spliced, the machine will find it difficult to drive over the joint. Many people buy the cassette machine as a substitute for the stereo record-player. These users would be well advised to invest in two good-sized loudspeakers, and a mid-priced recorder, than to pay a high price for a teak-and-stainless-steel unit coupled to tiny wooden boxes masquerading as loudspeaker cabinets.

Unlike the gramophone record, it is not possible to start playing a particular track on a music cassette at a moment's notice. But the cassette does have two sides to play, just as the LP record. As the tape runs from left to right, the heads 'read' the two stereo tracks recorded on the upper half of the tape. When the tape has run its length and the right-hand bobbin contains all the tape, the cassette can be taken from the machine and turned around so that the full bobbin is now on the left. When played, the heads will 'read' the other two stereo tracks laid on the other half of the tape. If the machine is fitted with a mono-head, then this will be twice the size of the separate stereo heads so that it reads and automatically combines the two tracks.

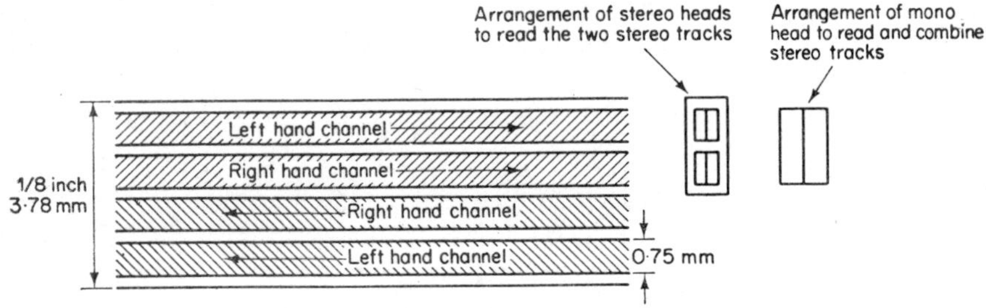

FIGURE 3—CASSETTE TAPE TRACKS

So that certain recordings can be found amongst the string of material on the tape, many machines are fitted with a tape indicator mounted immediately above the cassette deck. This consists of a numbered dial coupled to a hairline pointer that can move laterally across the cassette. When a particular recording on the tape is reached that you wish to replay, the machine is stopped and the dial rotated until the pointer is in line with the outside edge of the bobbin of tape visible through a window in the cassette casing. If the number on the dial is read off, then the recording can be refound by resetting the indicator to this number, and spooling the cassette until, once again, the tape bobbin is in line with the hair-line pointer. Although this seems far more complicated than setting the gramophone needle down on the appropriate track, with a little practice, the cassettee user can find any desired piece of music on the tape quite quickly. He will also be unable to scratch the recording as can be the case with traditional gramophone records.

A tape-recorder is a mechanical and electronic piece of apparatus, and often misused, and is likely at some time to go wrong. The cassette machine is transistorised, which gives its electronic circuitry long life and reliability. However, it can easily be damaged if the batteries are inserted the wrong way round, and care must be taken so that they are correctly connected. If the machine is dropped, the internal loudspeaker magnet may become removed from the loudspeaker coil. This may be remedied by using an external speaker in a cabinet with the recorder, which in any event will give much better reproduction than the speaker inside the machine's case. If the tape starts to run slow, the batteries may well need replacing. If the pitch of the recording varies up and down, the cause is more likely to be the cassette and not the recorder. The tape is probably unevenly wound inside the cassette and is feeding towards the head in a jerky manner. This is commonly caused by constantly spooling the cassette as opposed to allowing it to run its length, and the fault is frequently experienced with C120s.

Care should be taken not to force a cassette into a machine. There will be something solid blocking its path if the cassette does not easily slip in, and by forcing things the capstan may be bent, which will be difficult to realign properly. An investigation should relieve the blockage, but more often than not the user is simply trying to insert the cassette into the machine the wrong way round.

As with standard reel-to-reel recorders the arch enemy

is dirt. It is possible to buy head cleaning cassettes which, when run in the machine in place of a normal cassette, will clean and polish the heads. If there is dirt on the heads, the cassettes will sound woolly and indistinct, and since the head gaps are so narrow on cassette machines, the effect is more marked than with standard apparatus. It is quite possible to clean the heads, capstan and pinch-wheel with cotton buds and a little methylated spirits, but the special cleaning cassettes eliminate the fiddling and do an excellent job.

Some mains and battery machines are prone to add a hum, originating in the mains electricity supply, to any recordings made with the apparatus powered from the mains. The only solution is to use the battery supply when recording, and power the machine from the mains only during replay.

Figure 4 shows the Philips model 2205 mono cassette recorder, and the figures indicate the various facilities on the machine. The machine operates from both batteries and the mains, and, although mono, it will of course also reproduce stereo pre-recorded music tapes in a mono form.

1. This is the connector socket for remote control, which can also be used to provide a replay link to head-phones. This connector is usually wired to a switch on the microphone, which when thrown starts the machine motor. This remote on/off switch will work whether the machine is recording or playing back.

2. This connection socket takes the input plug from the microphone, radio, amplifier, record-player, another tape machine, or feed from a telephone. The connections are explained later in this chapter.

3. This is the lid that covers the cassette drive deck. With the lid up, the cassette can be slipped beneath it, and when the lid is closed the tape is fitted neatly around the capstan and drive spindles.

4. The tape-indicator dial linked to a hair-line pointer that moves across the window to the right of the dial.

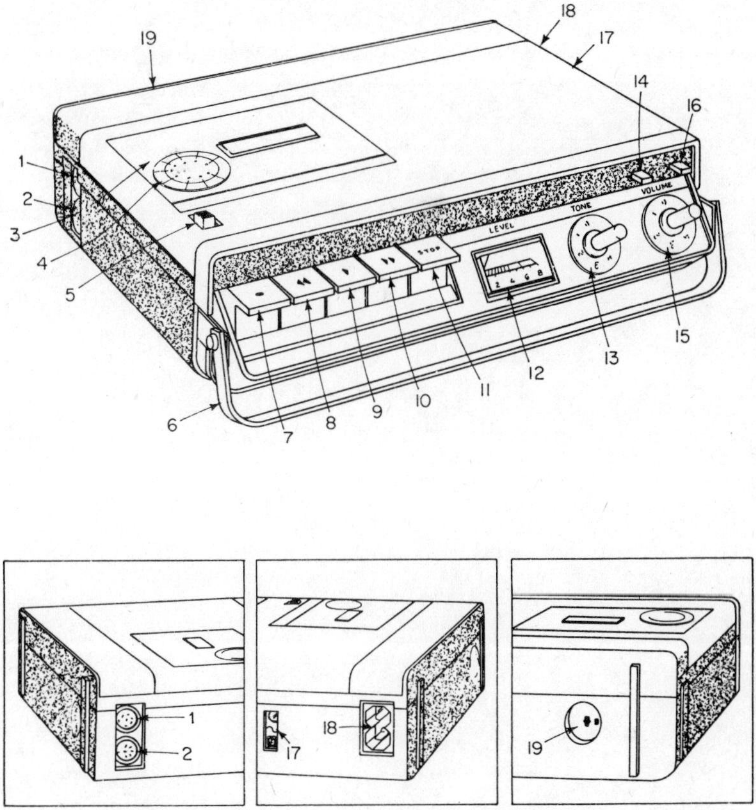

FIGURE 4

5. The cassette holder release knob which causes the lid above the cassette drive deck to pop up. To take out a cassette, the knob is pulled backwards and the cassette springs up with the lid.

6. A handle which doubles as a stand for the machine.

7. This is the record button. It must be depressed at the same time as the start button to record. If a remote start switch, probably on the microphone, is wired to the connector labelled (1) then this switch must be pressed to start recording. It is possible to adjust the record volume control (15) by depressing only the record button and

speaking into the microphone. The adjusted signal will be seen to register on the VU meter, although the machine is not recording.

8. The rewind button or fast backwards spooling as it is known. When this button is pressed, the tape rewinds on to the left-hand bobbin of the cassette.

9. The start button. With this button pressed and the volume control turned up, the machine will play back a cassette that has been previously recorded, such as a music cassette. If there is a remote control switch on the microphone, then this must also be pressed for the machine to work. The start button must also be pressed for recording, in conjunction with the record button.

10. This button winds the tape on to the right-hand bobbin of the cassette. It fast spools the tape forward so that a position later in the recording can be easily reached and replayed.

11. The stop button which will bring playback or recording to an end.

12. This is the recording-level indicator and battery-voltage indicator. On recording, the record volume control (15) should be so adjusted that the needle of the meter swings just up to the start of the red area on the indicator dial. If it swings *into* the red area, then the record volume control is set too high and distortion of the recording will result. If the meter needle does not swing up close to the red area, then the record volume control is set too low and the recording will be indistinct and hissy. On playback the meter shows the state of the machine's batteries. The needle should be in the green area of the indicator. If it lies in the red or white area, then the batteries have run too low and the machine will run at a slow speed. Replace the batteries: a fresh set will last about twenty hours if the machine is used for about two hours at a time.

13. The tone control. This should be adjusted to give a pleasant sound on playback. It does not affect the recording process at all.

14. This is a push-button to light up the recording-

level indicator when the machine is operating from its internal batteries.

15. The record level and replay level volume control. This one control has a double function; one in the replay mode, the other in recording. During recording, the control must be set so that the incoming signals from the microphone, radio or record-player are amplified to the correct amount to ensure a good tape. The position of the control will be indicated by the record volume meter. It should be adjusted so that the meter's needle just swings up to the red area of the dial. During playback the control is readjusted so that it gives a comfortable listening level. On very many machines this control will simply be a replay volume control only and the amplification for recording is adjusted automatically.

16. This switch changes the machine from battery to mains working. Naturally the mains cable must be attached to the machine and a power point for this switch to work.

17. The voltage selector. This screwdriver operated switch is to adjust the machine to work on the mains voltage in your area.

18. This is the connection socket for the mains lead. Some recorders have their mains lead wired directly into the recorder, in which case there will be no such socket. And of course very many recorders will only work from internal batteries.

19. This socket is designed to take a plug connected to a loudspeaker cabinet.

More sophisticated machines may well have separate record and replay volume controls, and extra tone controls to adjust the treble and bass independently. Stereo machines will have two VU meters, two record volume controls and two replay volume controls, one for each channel.There will also be two outputs: one for the left-hand speaker, and one for the right. Or the stereo machine may plug in to your main stereo amplifier system with a special interconnecting cable. As with mono machines, stereo cassette recorders can be fitted with an automatic recording device, in which case there will be no record volume controls.

Recent machines may be fitted with a switch marked Fe_2O_3 in one position, and CrO_2 in the other. These symbols stand for ferric oxide and chromium dioxide respectively. In the ferric oxide position, the machine is aligned to operate with the standard ferric oxide cassettes; while in the other position, the machine is switched for operation with the new chromium dioxide cassettes.

Your recorder may have a pause button or more likely a remote stop/start switch on the microphone. With this remote switch connected into the recorder, or by using the pause switch, and with the input of the machine connected to a record-player, it is quite easy to record music tapes without long pauses between the various songs. After the song has finished, the pause or remote off switch is pressed while the next disc is set up on the record-player. With the needle on the disc set a little back from the start of the song, the pause switch can be released, and the record played in. The disc may be held still while the turntable rotates beneath it by fitting a slip mat of felt between the disc and the turntable. Any rubber mat should be taken off the turntable first.

The cassette recorder is linked to other apparatus, an extension loudspeaker in a proper cabinet, and to its microphone by means of a multi-way cable and Din plugs. The major manufacturers have available interconnecting cables ready wired for linking practically any type of apparatus with another, including those with different sockets to the cassette machine. However, if you have a connecting lead supplied with the machine, it is also possible to re-adapt this to most purposes by examining its design and altering as required.

Figure 5 shows the standard arrangement for a three-pin cable, and the plug wiring diagrams below indicate how the cable should be fed. For example (Figure 6), since the red-coloured wire of the cable is connected to pin 3 of the input plug to the recorder, it is

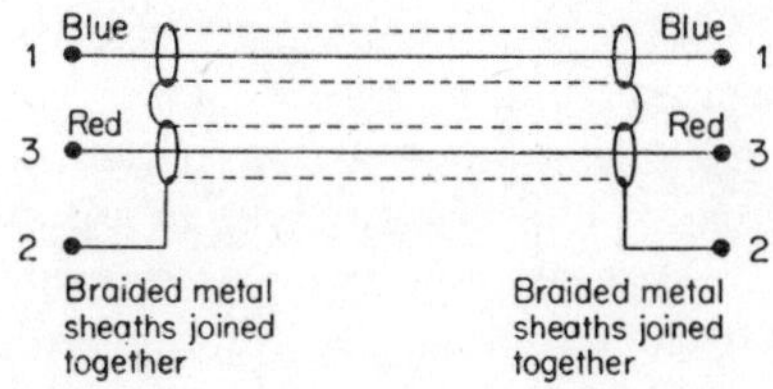

FIGURE 5—INTERCONNECTING CABLE WIRING

possible to connect this directly to one extension loudspeaker terminal of a radio, with the wire braided sheath of the cable connected to the other terminal. With this arrangement, the cassette machine can record from an older radio, but it is important to adjust the volume control of the wireless to a low position (but not off) and to turn the tone control to full treble.

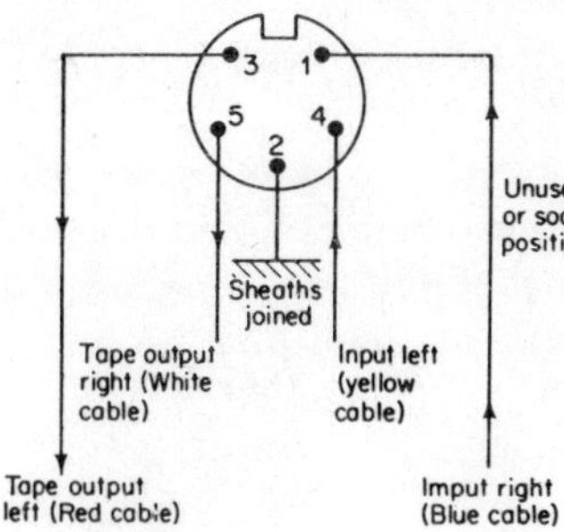

FIGURE 6
5-pin
stereo
connections
to external
amplifier

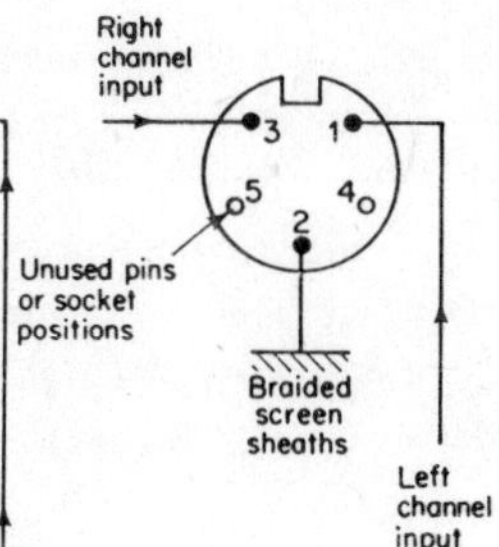

FIGURE 7
3-pin
stereo
connections

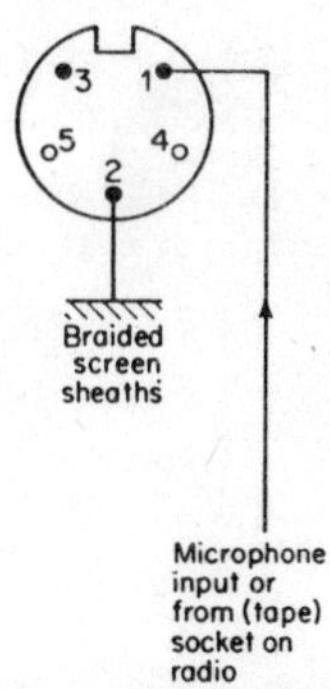

FIGURE 8
Mono
microphone
connections

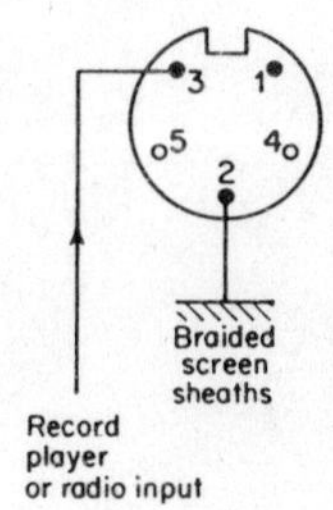

FIGURE 9
Mono
record-player
or radio
connections

The views of din socket connections are of the pin contacts looking at the soldering eyes of the plug, or into the holes of the socket. Although the diagrams 6—9 are 3-pin connections, often 5-pin sockets and plugs are used, with pins 4 and 5 left unconnected

Again, Figure 7 shows how a simple stereo record-player can be coupled to a stereo cassette player with a three-pin input socket. The first connector diagram illustrates how a stereo cassette player can be linked into a complete stereo outfit, recording from the stereo record-player, and replaying through the main amplifiers and loudspeakers. In this case there will be a four-way interconnecting cable, and the screening sheaths provide the fifth connection.

To record from the radio, record-player, or another tape-machine, three connection plugs and sockets are all that is required for mono or stereo. Although there are only three pins in use, it is not uncommon to find five-pin sockets catering for these input plugs. This may be confusing at first, but is useful because it means that five-pin Din plugs can be used, leaving pins four and five unconnected. If three-pin sockets are fitted to the machine by the manufacturers, this requires the use of three-pin plugs. If all you have available is a five-pin plug, it is always possible to lever out pins four and five with a pair of long-nose pliers.

Many recorders will have a separate socket to accommodate a second plug running from the microphone, besides the signal input lead. This second connector allows a switch on the microphone to stop and start the recorder's motor.

The Tape Cassette

The Compact Cassette contains the bobbins of tape that are used with the Cassette Recorder. Through the cassette there are two holes that coincide with the bobbin centres and, when placed into the machine, these bobbins lock on to the feed and take-up spindles that protrude into the cassette holes. To ensure a firm lock, each bobbin centre has teeth that fit into appropriate spaces in the tape-machine spindles. Between the two spindle holes on the cassette there is a window so that the tape can be seen. This enables the operator to judge

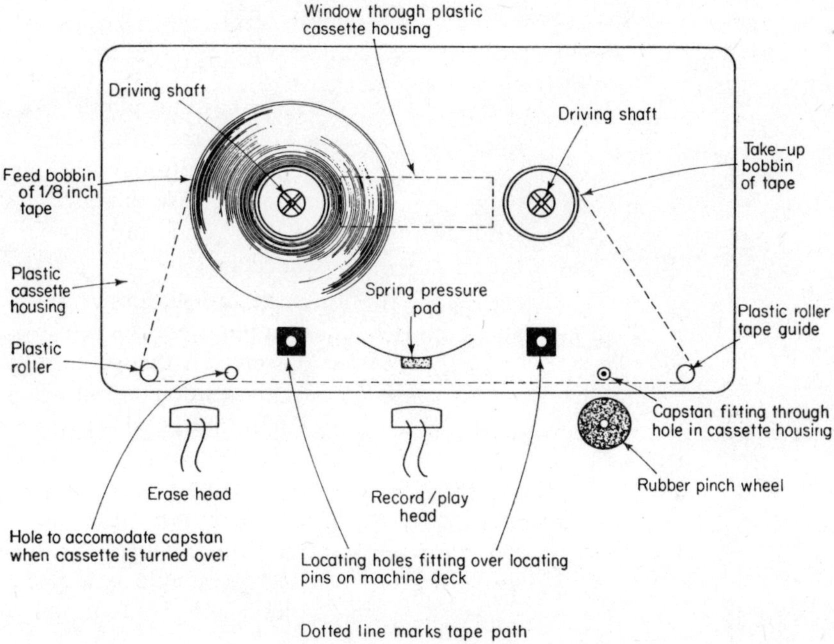

There is a gap in the plastic housing along the bottom side of this cassette diagram. When the machine is switched to play or record, the heads and the pinch-wheel are pushed slightly into the cassette and into contact with the tape. The pinch-wheel clasps the tape against the rotating capstan which pulls it at constant speed past the heads. The pressure pad holds the tape close to the replay/record head.

FIGURE 10—THE CASSETTE

whereabouts on the cassette he is playing or recording. It also provides a guide to adjust the tape-indicator on the machine.

On one side of the tape cassette there is a gap in the plastic case. Across this gap the tape is guided, running from one bobbin to the other. It is so designed that the coated magnetic side of the tape faces out of the gap, and this means that, unlike conventional machines, the tape is wound around the bobbins with the coated side outwards and shiny backing side inwards. In the centre of the gap in the cassette there is a *pressure-pad* mounted so that the tape passes across its padded face as it moves across the gap in the plastic. It is this pressure-pad that holds the tape closely in contact with the *head* of the machine when the cassette is in use.

To drive the tape across the machine heads at constant speed, the cassette package needs to allow for the rotating *capstan* and the *pinch-wheel* to clamp on to the tape. For this there are two holes through the plastic cassette, one at each end of the tape gap. These enable the machine's spindle-type capstan to fit immediately behind the tape to the right of the magnetic head, no matter which side of the cassette is being recorded or played back. When the recorder is switched to operate, the pinch-wheel clamps the tape firmly to the rotating capstan and the tape is pulled smoothly across the heads, from the left bobbin and on to the right.

There are normally two further holes through the cassette, just behind the tape gap, and these enable the tape-cassette to be firmly held in the machine by sliding over two tight-fitting metal prongs. The two plastic halves of the plastic casing may be held together around the tape bobbins by being welded around the sides, or may have a screw fastening through the centre of the box. This second type, although more expensive, does allow for decreasing the pressure on the bobbins if it is too great, and for refitting ends of the tape to their bobbin should they break apart.

On the market at this present time there are many different types of tape cassettes available, and it is important to be clear about their differences and their uses. Firstly there are the pre-recorded cassettes. These are manufactured by the commercial gramophone companies and contain the same material as the traditional long-playing album record. They generally play for twenty minutes a side, although the tape is normally longer, leaving a spare length of 'blank' at the end of each side. There are 'double-play' cassettes that contain the contents of two LPs. Naturally the cost of these *music cassettes* is high, since you are paying for the recorded work and not simply the tape package. These cassettes, therefore, should never be used for recording, since you will be destroying the music for which you have paid so dear. Blank cassettes are easily obtainable for recording, at a very modest price.

The standard blank cassette is the C60, which will play for thirty minutes per side. Then there are the C90 and C120 cassettes that play for forty-five and sixty minutes per side, respectively. Recently a fifteen minute sided C30 cassette has been introduced to fill the demand from letter-tape correspondents for a shorter and lighter package.

Each of these blank cassettes is available with standard tape coated with ferric oxide, or with low-noise chromium dioxide tape. The decision of which type of tape to buy will depend upon the quality of your cassette recorder and the price you wish to pay. It is not possible simply to pay the extra cost of the chromium dioxide cassette and expect superior recordings from a cheap machine. Indeed the normal basic recorder will give inferior recordings with this new low-noise tape, since its *bias* level has been adjusted to suit the standard ferric oxide tapes. However, with a more sophisticated recorder it is possible to throw a switch to suit the type of tape being used. With such a machine it then becomes possible to buy cheap standard tape cassettes for corresponding or perhaps dictation purposes, while spending the extra for chromium dioxide cassettes for recording music, when a higher quality is expected.

For most recording purposes, the C60 cassette is most convenient, since it has a reasonable playing time, yet is not so long as to conceal any part of the recording that may wish to be played and replayed. For music and party tapes, the C90 and C120 can be recorded to provide long-playing runs of music, and would therefore be the choice whenever the whole cassette is required to run from start to finish.

It would appear that the C120 with its extended

playing time is ideal for compiling music tapes recorded from the radio or from a record player. However, the C120 uses a thinner tape than the C60 or C90 and, because of its low transverse rigidity, it is prone to accident. On starting the tape it may be stretched into a bump which causes the winding on the bobbin to become uneven. On replaying an unevenly wound cassette, the recording will sound unsatisfactory owing to the wow caused in the machine. The thin tape may also slip from the coil once it has been wound on to the bobbin, which again will affect the smooth action required for good reproduction. These problems arise with all cassettes, but they are more common with the C120, and if the cassette is to be constantly stopped, restarted and rewound it is better to choose the C90 with its slightly thicker tape.

To avoid bad winding and buckling of the tape in any cassette, care should be taken to store the recordings in their correct dust-free cases. Before using a cassette, ensure that the bobbins are closely coiled with the tape. If free loops of tape can be seen through the window in the plastic cassette, then a pencil should be pushed into the hole at the centre of one of the bobbins and gently rotated until the tape is firmly wound. Those cassettes that are kept together with a screw fastening may be clamped too tight so that the tape is kept from running freely. The obvious solution is to release the screw slightly so that the bobbins rotate without friction.

The cassette recorder has provided packaged loading of the medium and in so doing has taken the operator's mind away from the fact that it is still tape *magnetic recording*. As such, the system is open to all the problems of normal tape-recording, and since the tape speed is so low, and the tape far more fragile than normal recording tape, a slight fault is immediately noticeable. If a recorded cassette is placed near a magnetic field, it will be affected, usually partially erased. This means that cassettes must never be placed on or near any electrical device, and this very definitely includes loudspeakers!

Dirt and dust will ruin the machine and the recordings, and great care should be exercised when taking the recorder to the beach, as a few grains of sand can sabotage cassettes and the moving parts of the machine. Similarly, just a drop of liquid spilt on the tapes can bring disaster. As mentioned earlier, there are special head-cleaning cassettes available which will remove dirt from the head faces and this should be carried out regularly.

To solve the problems of uneven reeling and slipping of the tape from the coil within the cassette, a new addition has been incorporated into the tape guiding system within the plastic casing. In these *Special Mechanics* cassettes (or SM cassettes) there are two plastic tusks and a spring ensuring that the tape is wound neatly and stays in place. The tusks are mounted inside the cassette in such a way as to move sideways to accommodate changes in coil diameter as the tape moves from one bobbin to the other. However, they are arranged so that they cannot move up or down. Each tusk is moulded with a shallow 'U' shape, which guides the tape precisely on to the take-up bobbin when the machine is running. Only one tusk is in action at any one time; the other simply acts as a gentle guide for the tape leaving the feed bobbin. When the cassette is turned around for playing or recording on the other half of the tape, the roles of the two tusks are reversed. The active tusk at any time is pressing gently on to the take-up coil of tape by the tension of the moving tape itself, and this section transfers the moving tape on to the coil at a constant level. The result is an evenly coiled tape.

Once the tape is wound on to the already coiled tape, it can still slip down the winding, particularly if it has the low transverse rigidity of the thinnest tapes used in the C120. The Special Mechanics cassette overcomes this tendency by inserting a plastic spring clip which applies gentle pressure to both coils at the spot where slipping is likely to take place.

The Special Mechanics cassette was developed in

Germany by BASF, although there are now similar types by other manufacturers. They are available in all durations of playing time, and are fitted with either standard or chromium dioxide tapes. They are particularly useful when a long playing time is required, since it is the C120 that suffers most from the problems that these special mechanics were designed to solve.

Chromium Dioxide Tape

Since the cassette recorder operates at the low speed of 1⅞" per second, its response to high treble frequencies is not very satisfactory. To improve this performance, the high frequencies are boosted prior to recording. However, besides amplifying the high notes, this also amplifies the background noise, so that when using normal tape cassettes the system is inferior to that of a reel-to-reel recorder running at a faster speed. It is impractical to increase the speed of the cassette machine as this would limit the recording time available on the compact cassette and produce extra difficulties in the smooth winding of the narrow tape on to the bobbin spools.

There have been two revolutions in tape-recording that now enable the cassette recorder to rival the reel-to-reel recorder in terms of high quality, although these developments have been equally applied to conventional machines. The arrival of chromium dioxide tape and the development of the Dolby Noise Reduction System are responsible for allowing high-quality recording systems to use the Compact Cassette.

A cassette using chromium dioxide tape has as advantages over the standard ferric oxide tape: improved frequency response, improved background noise at high frequencies, and improved high-frequency dynamic range. This enables reproduction to include those frequencies that give the brilliance, whereas we have all experienced the tendency towards woolly lifeless recordings of music with standard cassettes. Chromium dioxide tapes produce the same recording quality sound as conventional tape running at twice the speed. Therefore, a cassette recorder can parallel the performance of a reel-to-reel recorder operating at 3¾″ per second with standard tape. Similarly, a reel-to-reel machine running at 3¾″ per second and using chromium dioxide tape will perform as well as would be expected at 7½″ per second with ordinary tape.

However, the newer chromium dioxide tape requires a high level of erase current to clean the tape, and a higher level of bias current than normally used. In older machines, domestic recorders, and certainly with the cheap range of cassette recorders, it is not possible to adjust the erase or bias currents and so this tape is unsuitable. However, with more sophisticated units, it *is* possible to adjust to the standard tape or CrO_2.

Many older quality cassette machines *can* in practice be used with chromium dioxide cassettes. The bias setting, although not set for the new tape, is usually high enough. With normal tape the recorder needs to boost the treble frequencies and this equalisation setting will cause the chromium dioxide recording to contain high frequency over-emphasis. However, if the machine is fitted with a replay tone control, this can be adjusted so that the cassette on playback sounds more balanced. On the other hand, should you wish to erase the cassette and re-record with different material, the unadjusted recorder cannot cope. The standard recorder is not capable of erasing a chromium dioxide cassette and so a new cassette must be used on each occasion unless a new machine, designed for this tape, is used.

The Dolby System

The Dolby B system of noise reduction consists of two units, one of which is in the recording circuit and the other in the replay. On record, the Dolby unit amplifies all the low-level treble signals so that they are all at the same high level. During the recording process the noise of the system is also recorded and this high-frequency noise is particularly troublesome appearing as ‘hiss’.

However, since the Dolby unit has amplified all the low-level high-frequency *signals,* during playback these self-same signals must be equally attenuated for the recording to sound natural. During this attenuation in the Dolby replay unit, the hiss of the system is also attenuated by the same amount. So the two Dolby units are capable of reducing noise in the tape-recording process. But it should be clearly understood that the Dolby system cannot clean up a previously recorded noisy tape, it can only improve the performance of a tape machine. This is particularly useful with cassette recorders, since they tend to boost the treble signals to overcome their loss of frequency-response owing to their slow tape speed. In boosting the treble, they naturally also boost the hiss content of the system. Using Dolby amplifiers, and the high-frequency sensitive chromium dioxide tape, cassette machines can now operate to professional recording standard.

The Dolby A system is used in professional music recording studios and reduces noise in the system throughout the complete audio range. Where the Dolby B system just treats the treble frequencies, the Dolby A splits the sound to be recorded into four separate frequency bands. Each band is then treated in the same manner as explained with the Dolby B system. The four treated frequency bands are then recombined during replay.

It is certainly not worth considering the finesse of such noise reduction units if you are content to listen to music through the speaker in the cassette recorder's case. However, these units are important should you consider reproducing cassettes for distribution, when every copy increases the noise on the tape. In these days when students' lectures, talking books for the blind, and even sermons are being recorded and distributed by copying on to cassettes, the Dolby B unit is enabling the sound quality of the copies to resemble closely the original recorded sound.

The Radio Studio

Since local radio is springing up all over the country and more and more people are being involved in broadcasting, even if it is only to be interviewed on the street, I thought I would add a note at the end of this book on the radio studio. It is easy to be dazzled by the array of stainless steel and teak equipment in the studio, but in essence everything that is there you may have used at home, albeit in a different form.

The hub of the studio is the sound mixer, where all the various components that build up a radio programme are fed together in the correct order and in the correct volume. On the mixer are the channel faders, which are simply volume controls connected to the various pieces of apparatus that feed signals into the programme. There will be a record-player channel, a replay tape machine channel, a microphone channel fader, a cartridge machine channel, and so on. So that the technician operating the mixer desk can judge the volumes of all the contributing signals, there is a loudspeaker so he can *hear* the result, and a meter so that he has a visual indication. There is also a facility called 'prefade' or 'audition' on every channel. This enables the operator to hear a particular channel, identify the tape or start of a disc, and measure its volume, *without* the channel fader being open. In this manner the technician can set up records about to be played and check that the tape about to be included is the correct one. On each channel there are usually one or two tone controls. These enable the sound quality of all the bits of the programme to be matched. A tape recorded on location may sound a little woolly and would have its treble boosted, while an old 78 record will be treated with treble cut to decrease the hiss and scratch.

In many modern radio stations the mixer console operator is also the programme presenter, so that the production can be fast and smooth. It is in this situation that it becomes clear that good artistic sound mixing relies not so much on the equipment as on the person operating it. The disc jockey is also able to use his voice to match and cue the sounds that he brings in from disc, tape and telephone line. This is indeed technical artistry, and a person with a flair for sound, a touch of acting ability and a thorough working knowledge of his electronic sound equipment, can produce on simple equipment far better entertainment than a whole production team in a vastly expensive studio. If you enjoy tape-recording and creating sound productions, perhaps you would benefit by joining a Hospital Broadcasting Unit to learn radio techniques.

Try telephoning your local large hospital and asking to be put in touch with the person in charge of Hospital Broadcasting. This person will usually be only too glad to be offered extra voluntary assistance, and you will be given opportunities to mount as complicated a programme as you can handle.

Most colleges and universities have recording facilities and even campus radio stations. These again are an excellent training ground in radio production and offer the chance to use expensive electronic equipment. Speak to the president of your local students' union or write to the Association of Campus Radio Stations, University Radio Loughborough, Loughborough.

It is only a small hop from one of these amateur studios to working for a BBC or commercial local radio station. But beware of the many private broadcasting and dee-jay schools that advertise in the press. They are expensive and do not enjoy a high reputation.